OPPORTUNITIES TO USE REMOTE SENSING IN UNDERSTANDING Permafrost AND RELATED ECOLOGICAL CHARACTERISTICS

Report of a Workshop

Committee on Opportunities to Use Remote Sensing in Understanding Permafrost and Ecosystems: A Workshop

Polar Research Board

Division on Earth and Life Studies

NATIONAL RESEARCH COUNCIL
OF THE NATIONAL ACADEMIES

THE NATIONAL ACADEMIES PRESS
Washington, D.C.
www.nap.edu

THE NATIONAL ACADEMIES PRESS • 500 Fifth Street, NW • Washington, DC 20001

NOTICE: The project that is the subject of this report was approved by the Governing Board of the National Research Council, whose members are drawn from the councils of the National Academy of Sciences, the National Academy of Engineering, and the Institute of Medicine. The members of the committee responsible for the report were chosen for their special competences and with regard for appropriate balance.

This study was supported by the National Aeronautics and Space Administration under contract number NNX13AD79G. Any opinions, findings, and conclusions, or recommendations expressed in this material are those of the author(s) and do not necessarily reflect the views of the sponsoring agencies or any of their subagencies.

International Standard Book Number-13: 978-0-309-30121-3
International Standard Book Number-10: 0-309-30121-1

Additional copies of this report are available for sale from the National Academies Press, 500 Fifth Street, NW, Keck 360, Washington, DC 20001; (800) 624-6242 or (202) 334-3313; http://www.nap.edu/.

Printed in the United States of America

THE NATIONAL ACADEMIES

Advisers to the Nation on Science, Engineering, and Medicine

The **National Academy of Sciences** is a private, nonprofit, self-perpetuating society of distinguished scholars engaged in scientific and engineering research, dedicated to the furtherance of science and technology and to their use for the general welfare. Upon the authority of the charter granted to it by the Congress in 1863, the Academy has a mandate that requires it to advise the federal government on scientific and technical matters. Dr. Ralph J. Cicerone is president of the National Academy of Sciences.

The **National Academy of Engineering** was established in 1964, under the charter of the National Academy of Sciences, as a parallel organization of outstanding engineers. It is autonomous in its administration and in the selection of its members, sharing with the National Academy of Sciences the responsibility for advising the federal government. The National Academy of Engineering also sponsors engineering programs aimed at meeting national needs, encourages education and research, and recognizes the superior achievements of engineers. Dr. C. D. Mote, Jr., is president of the National Academy of Engineering.

The **Institute of Medicine** was established in 1970 by the National Academy of Sciences to secure the services of eminent members of appropriate professions in the examination of policy matters pertaining to the health of the public. The Institute acts under the responsibility given to the National Academy of Sciences by its congressional charter to be an adviser to the federal government and, upon its own initiative, to identify issues of medical care, research, and education. Dr. Harvey V. Fineberg is president of the Institute of Medicine.

The **National Research Council** was organized by the National Academy of Sciences in 1916 to associate the broad community of science and technology with the Academy's purposes of furthering knowledge and advising the federal government. Functioning in accordance with general policies determined by the Academy, the Council has become the principal operating agency of both the National Academy of Sciences and the National Academy of Engineering in providing services to the government, the public, and the scientific and engineering communities. The Council is administered jointly by both Academies and the Institute of Medicine. Dr. Ralph J. Cicerone and Dr. C. D. Mote, Jr., are chair and vice chair, respectively, of the National Research Council.

www.national-academies.org

COMMITTEE ON OPPORTUNITIES TO USE REMOTE SENSING IN UNDERSTANDING PERMAFROST AND ECOSYSTEMS: A WORKSHOP

PRASAD GOGINENI (*Co-Chair*), University of Kansas, Lawrence
VLADIMIR E. ROMANOVSKY (*Co-Chair*), University of Alaska, Fairbanks
JESSICA CHERRY, University of Alaska, Fairbanks
CLAUDE DUGUAY, University of Waterloo, Ontario, Canada
SCOTT GOETZ, Woods Hole Research Center, Falmouth, MA
M. TORRE JORGENSON, Alaska Ecoscience, Fairbanks
MAHTA MOGHADDAM, University of Southern California, Los Angeles

NRC Staff

KATIE THOMAS, Program Officer
LAUREN BROWN, Associate Program Officer
SHELLY FREELAND, Senior Program Assistant

Acknowledgments

This workshop summary has been reviewed in draft form by persons chosen for their diverse perspectives and technical expertise in accordance with procedures approved by the National Research Council's Report Review Committee. The purposes of this review are to provide candid and critical comments that will assist the institution in making the published summary as sound as possible and to ensure that the summary meets institutional standards of objectivity, evidence, and responsiveness to the study charge. The review comments and draft manuscript remain confidential to protect the integrity of the deliberative process. We wish to thank the following for their participation in the review of this summary:

GUIDO GROSSE, Alfred Wegener Institute, Potsdam, Germany
LARRY D. HINZMAN, University of Alaska, Fairbanks
ANNE W. NOLIN, Oregon State University, Corvallis
EDWARD SCHUUR, University of Florida, Gainesville

Although the reviewers listed above have provided many constructive comments and suggestions, they were not asked to endorse, nor did they see the final draft of the workshop summary before its release. The review of this summary was overseen by **Jeff Dozier,** University of California, Santa Barbara. Appointed by the National Research Council, he was responsible for making certain that an independent examination of this summary was carried out in accordance with institutional procedures and that all review comments were carefully considered. Responsibility for the final content of this summary rests entirely with the author and the National Research Council.

Contents

Overview

Climate change is causing widespread thawing and degradation of permafrost, which has associated impacts on infrastructure, ecosystems, and the global carbon cycle. Data are needed to observe and monitor permafrost and for input into models that project permafrost change. Permafrost is a challenge to study because it is a subsurface condition of the ground, largely found in remote locations, and vastly distributed. An ad hoc committee of experts, under the auspices of the National Research Council, organized a workshop to explore opportunities for harnessing remote sensing technologies to advance our understanding of permafrost status and trends and the impacts of permafrost change.

Many workshop discussions focused on using remote sensing technologies to measure various permafrost properties and processes and other ecological characteristics that can be used to achieve better understanding of permafrost and its dynamics. Measurements of permafrost-related ecological variables provide some crucial information about changes in the relevant ecological characteristics and are used to extract information about permafrost conditions and processes. One innovative example of a permafrost-related ecological variable is the measurement of changes in seasonal micro-topography to estimate ice content in permafrost. Permafrost properties are those characteristics that are inherent to permafrost. Examples include ice content, maximum depth of seasonal thaw (depth to the surface of permafrost), and permafrost temperature. Currently, there are considerably more permafrost-related ecological properties that can be observed with remote sensing methods than permafrost processes and properties. Of the more than 60 permafrost and permafrost-related ecological properties and processes that were discussed during the workshop, the following emerged as having the most impact in advancing the current state of knowledge of permafrost landscapes, if they could be measured through remote sensing:

- Active layer thickness
- Ground ice (volume and morphology)
- Snow characteristics (extent, water equivalent, depth, density, conductivity)
- Surface topography (static, macro-, and micro-)
- Longer-term surface subsidence
- Thermokarst distribution
- Surface water bodies (including dynamics, redistribution)
- Surficial geology-terrain units (including lithology, bedrock)
- Soil organic layer (thickness, moisture, conductivity)
- Land cover (including spectral vegetation indices)
- Vegetation structure and composition
- Methane (flux or concentration)
- Water vapor flux
- Carbon dioxide (flux or concentration)
- Land surface temperature
- Subsurface soil temperature
- Seasonal heave/subsidence
- Soil moisture
- Biomass (above ground)

It was clear from the workshop discussions that innovative, multiscale, multisensor approaches, using ground-based, aircraft, and spaceborne instruments, would substantially advance the current state of knowledge of permafrost landscapes and, in the process, would provide critically needed information on subsurface properties that determine the vulnerability of permafrost systems to warming. The participants discussed the utility of remote sensing observations, both from existing sensors and those expected in the near future, and that it could be advanced through synergistic approaches that would permit derivation of data products characterizing critical permafrost properties across spatial scales. Advancement of techniques and algorithms was emphasized as a means to integrate field measurement with remote sensing observations, allowing improved direct and indirect retrieval of permafrost properties and thereby establishing a baseline against which change and longer-term trends can be assessed. The algorithms, data sets, and derived products would ultimately allow better model-data assimilation, initialization, and parameterization, and thus the advancement of more realistic permafrost models than would otherwise be possible. Taken together, field measurements, remote sensing–derived maps of properties, and improved models would advance our understanding and prediction of the state of permafrost landscapes and associated feedbacks to the climate system.

1

Introduction

PERMAFROST CHARACTERISTICS AND CHANGE

The most common and accepted definition of permafrost refers to the thermal conditions of the subsurface earth. Permafrost is defined as soil, rock, and any other subsurface earth material that exists at or below 0°C for two or more consecutive years (van Everdingen, 2005). This definition does not require the presence of ice. Because of this, the physical and biogeochemical properties of permafrost may vary widely depending on the characteristics of the parent material, ice and liquid water content, topography, biota, and climate. The complexity of permafrost evolution stems from its strong and nonlinear dependence on these ecological characteristics (Figure 1.1). Changes in permafrost are usually more complex than just a simple sum of changes induced by the variations in individual components of the environment. The resulting evolution of permafrost is usually even more complex because of the numerous reciprocal changes in other ecological components. Numerous positive and negative feedbacks exist between changing permafrost and other components of the Earth system, which make projections about permafrost change difficult. Among the most important impacts of changing permafrost environments are surface and subsurface hydrology, biota, biogeochemical processes in general, and the global carbon cycle in particular.

The formation, persistence, and disappearance of permafrost are highly dependent on climate because permafrost is a thermal condition. General circulation models project that a doubling of atmospheric concentration of carbon dioxide will result in mean annual air temperatures to rise up to several degrees over much of the Arctic. Discontinuous permafrost will likely ultimately disappear as a result of ground thermal changes associated with global climate warming, because ground temperatures in these regions are within 1-2 degrees of thawing. Permafrost degradation will have associated physical impacts where ground ice contents are high. Soils with the potential for instability upon thaw (thaw settlement, creep, or slope failure) may have significant impacts on the landscape (e.g., coastal erosion), ecosystems, and infrastructure.

Results from an international monitoring network of permafrost temperature measurements in boreholes (Thermal State of Permafrost) established by the permafrost research community clearly show warming of Northern Hemisphere permafrost in most regions as well as degradation in discontinuous permafrost regions during the last three decades (Romanovsky et al., 2010). In addition to impacts on northern hydrology and ecosystem characteristics (Hinzman et al., 2005), thaw also allows decomposition of sequestered organic matter, releasing currently stored soil carbon in the form of greenhouse gases (e.g., carbon dioxide and methane) to the atmosphere or to the hydrosphere (as dissolved and particulate organic carbon).

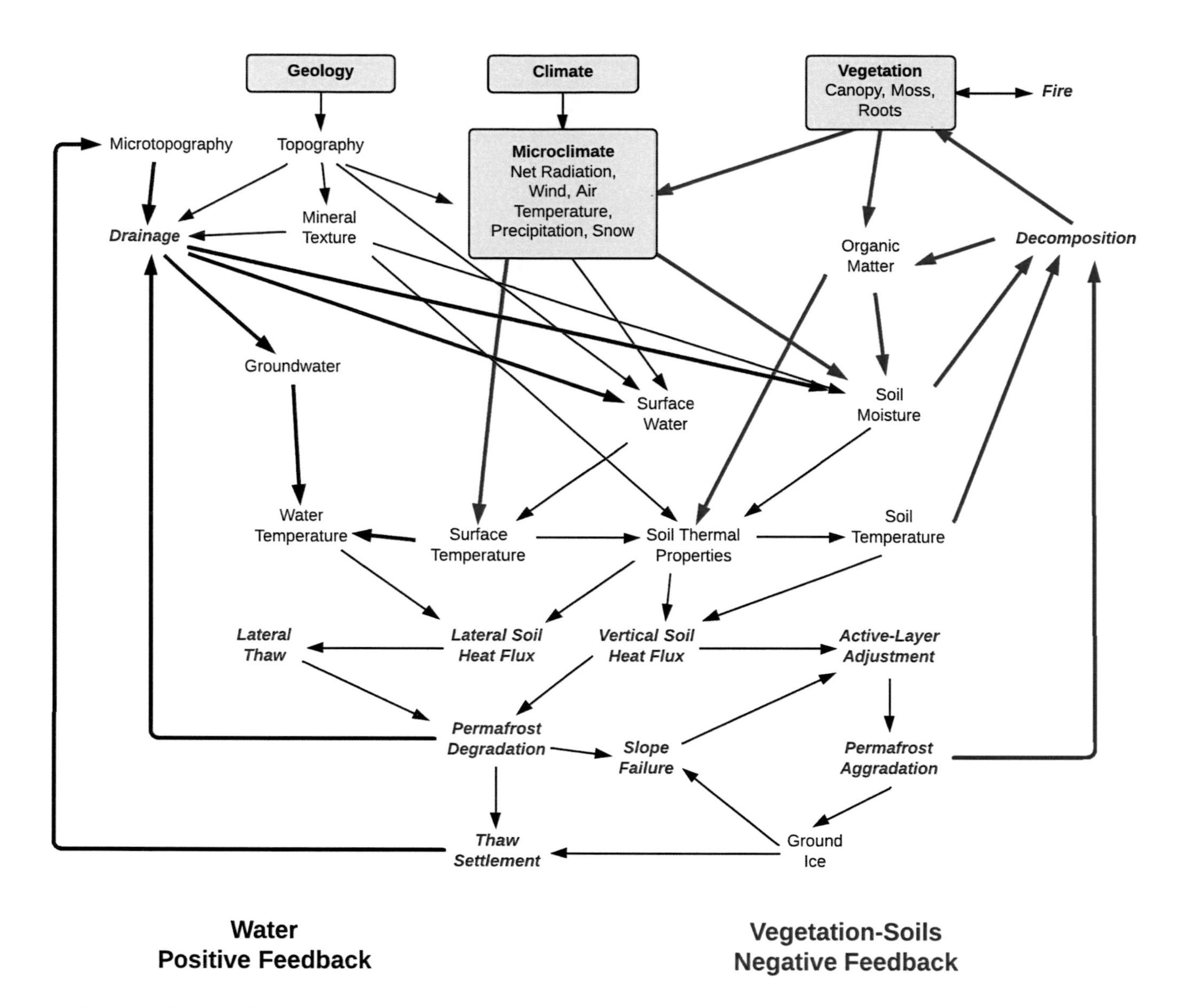

FIGURE 1.1 Conceptual diagram of ecological factors and feedbacks affecting permafrost aggradation and degradation that illustrates the complex interactions affecting the response of permafrost to surface boundary conditions and active-layer properties (Jorgenson et al., 2010). The text denotes properties (black text) and processes (blue text) that can be measured through remote sensing or field observations, while arrows indicate interactions among properties and processes. Black arrows indicate water positive feedbacks; green arrows indicate vegetation-soils negative feedbacks. Because permafrost properties, such as soil temperatures and ground ice, are difficult to measure through remote sensing, measurement of multiple surface properties and modeling will be needed to determine permafrost characteristics. SOURCE: Adapted from Jorgenson et al., 2010. © 2008 Canadian Science Publishing. Reproduced with permission.

DIRECT AND INDIRECT REMOTE SENSING OF PERMAFROST AND PERMAFROST-RELATED ECOLOGICAL CHARACTERISTICS

Permafrost thaw has wide-ranging impacts, such as erosion of riverbanks and coastlines, destabilization of infrastructure, and potential implications for ecosystems in the high latitudes, hydrology, and the carbon cycle. Data are needed to provide information on permafrost landscapes and on subsurface properties that determine the vulnerability of permafrost systems to warming. However, it is difficult to make in situ measurements of permafrost because of to the remoteness and vast distribution of permafrost around the globe (Figure 1.2). The temporal evolution of permafrost triggers change in many other ecological characteristics, such as variations in micro-topography, local hydrology, and vegetation. Such changes may be observed remotely from space or from an aircraft using various sensors with high spatial resolution. The relationship between permafrost and other ecological characteristics provides an opportunity to observe and document changes in permafrost that otherwise would be difficult to detect. Thus, remote sensing techniques can be applied to observe and monitor permafrost using indirect indicators of permafrost evolution.

In this report, we discuss two types of variables that can be observed with remote sensing to study permafrost. "Permafrost-related ecological variables" refer to those properties that can be measured to provide some crucial information about changes in the relevant ecological characteristics and to extract information about permafrost conditions and processes. These variables support understanding and modeling of permafrost properties or permafrost changes. An example of a per-

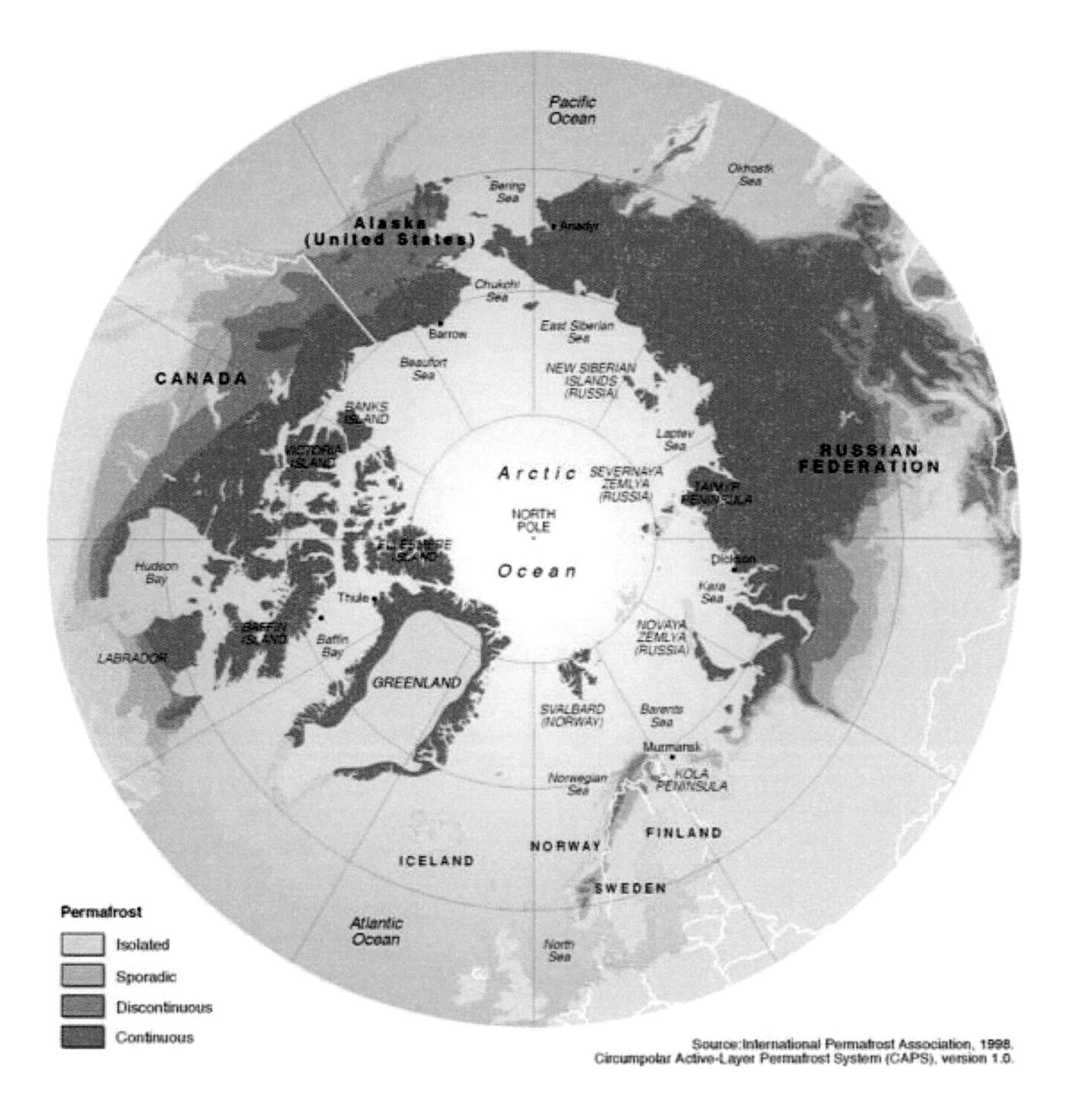

FIGURE 1.2 The distribution of permafrost in the Arctic. An updated circumpolar permafrost distribution map will be released once each Arctic country produces its own updated permafrost map. SOURCE: Philippe Rekacewicz, UNEP/GRID-Arendal (http://www.grida.no/graphicslib/detail/permafrost-distribution-in-the-arctic_3823).

mafrost-related ecological variable is the use of surface skin temperature to calculate the active layer thickness.

"Permafrost properties" refer to the essential key defining state variables of permafrost that then impact the ecological variables. These permafrost properties include

1. Ground temperature
2. Thickness of the active layer or the depth to the surface of permafrost
3. Thickness of permafrost
4. Spatial patchiness of permafrost
5. Ice content in permafrost

These properties describe where permafrost is, and what it is made of. Some of the variables were discussed in more detail at the workshop than others, which is reflected in the report. At present, considerably more ecological variables can be observed with remote sensing methods than can permafrost properties.

Any of the remote sensing methods, as well as any ground-based geophysical measurements, are indirect in comparison to truly direct methods of investigation such as drilling, in situ ground temperature measurements, and laboratory testing of samples collected in the field. Two types of approaches were discussed in the workshop for measuring both permafrost and ecological variables. "Direct" remote sensing methods can provide information about a variable of interest, such as ice content and permafrost temperature. "Indirect" methods incorporate modeling and remote sensing observations to estimate some crucial information about changes in the variable of interest (e.g., an ecological variable or a permafrost variable). One good example of an "indirect" method is the measurement of the evolution in micro-topography to estimate ice content in permafrost.

The Cryosphere Theme Report to the Integrated Global Observing Strategy (IGOS) partnership urges the development of remote sensing techniques to observe permafrost indirectly through connecting the observable land surface properties with subsurface permafrost characteristics (IGOS, 2007). The National Research Council (NRC) has recommended (NRC, 2007, p. 260) that "inferences be drawn from in situ measurements and remotely sensed observations from satellite and suborbital platforms." To date, however, no strategy or NASA missions *specifically address* the scientific questions surrounding permafrost degradation (NRC, 2007, Table 9.A.1).

The most direct indicators of changes in permafrost are its temperature and the active layer thickness (ALT). ALT is the thickness of the top layer of soil and/or rock that thaws during the summer and freezes again during the following winter. Permafrost temperature is best used as an indicator of long-term change at a depth where seasonal variations in ground temperature cease to exist. This depth varies from a few meters in warm, ice-rich permafrost to 20 meters and more in cold permafrost and bedrock (Romanovsky et al., 2010; Smith et al., 2010;). In addition, permafrost temperature is the best indicator of its stability. The closer this temperature is to the 0°C threshold, the less stable is the permafrost. Crossing this threshold triggers a widespread thawing of near-surface permafrost with negative consequences to the environment and infrastructure (Grosse et al., 2012; Instanes and Anisimov, 2008; Larsen et al., 2008).

Increases in the ALT may also trigger changes in other components of the system, even when permafrost is thermally stable with temperatures below 0°C. This may happen in areas with a large amount of ground ice. As soon as the summer thaw reaches the top of a very ice-rich layer or a pure-ice layer embedded in permafrost, any additional thaw triggers ground-surface subsidence. Because the distribution of ground ice is usually spatially uneven, the subsidence will develop a localized depression. Additional snow will collect during the winter and surface water will accumulate during the summer in this depression. Both of these processes will make the ground below the depression even warmer, and the local thawing of near-surface permafrost will progress more rapidly, developing into larger depressions, ponds, and eventually lakes. This process is called thermokarst formation and is typical within areas of degrading ice-rich permafrost. Development of thermokarst depressions and lakes will change local hydrology and have an impact on biota and trace gas emissions (Figure 1.3). This example highlights the importance of the volume and the morphology of the ground ice distribution as characteristics of permafrost. The impact of thawing permafrost on any infrastructure built above it also will strongly depend on these characteristics.

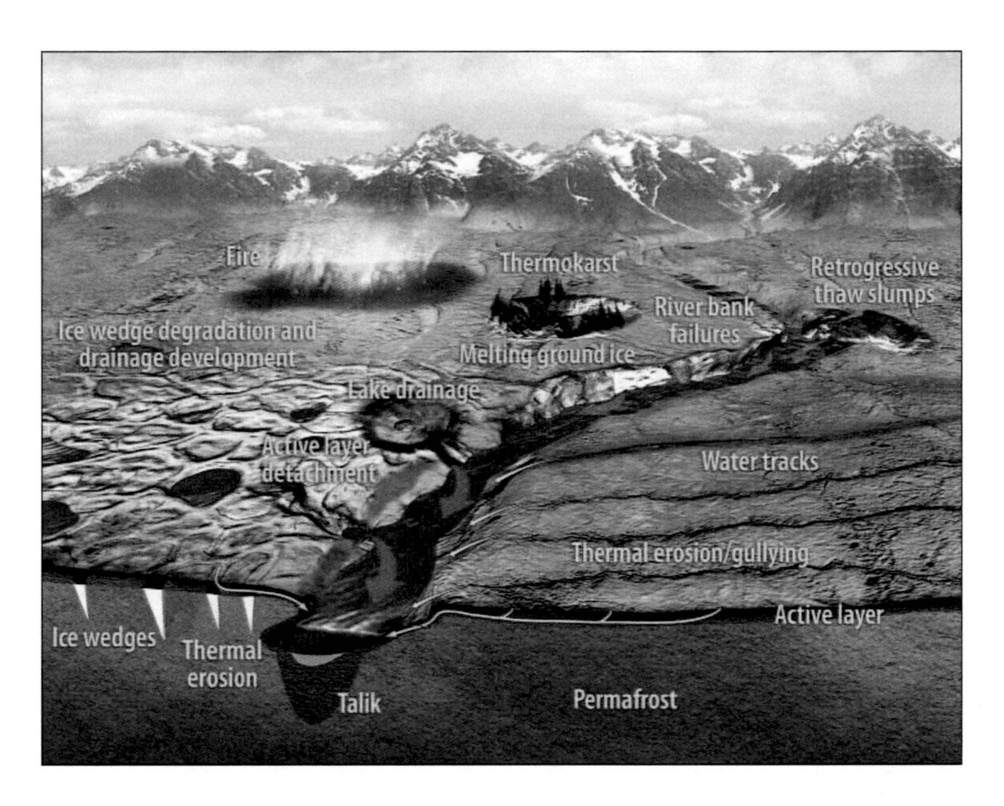

FIGURE 1.3 A schematic that illustrates the complex processes in a changing permafrost environment. SOURCE: Rowland et al., 2010.

U.S. AND INTERNATIONAL EFFORTS TO USE REMOTE SENSING TO STUDY PERMAFROST

Permafrost has been identified as an Essential Climate Variable in the Global Climate Observing System (WMO, 2010). Several national and international efforts have emerged in recent years to address how remote sensing can be used to study permafrost processes and change.

The planning committee was charged to recognize relevant efforts by NASA and other U.S. agencies. One key, emerging activity is ABoVE (the Arctic-Boreal Vulnerability Experiment),[1] which will take place in Alaska and Canada during the next 6 to 9 years. ABoVE is planned as an international research initiative led by NASA to produce new knowledge needed to understand how climate change affects ecosystems in the High Northern Latitude region and how these changes produce feedbacks to climate and are influencing ecosystem services. ABoVE will acquire, process, integrate, and synthesize geospatial information products generated from a combination of airborne and spaceborne remote sensing observations with data from field studies and ground-based monitoring to address six primary research objectives: (1) impacts and responses of human societies to environmental change, (2) changes in disturbance regimes and their impacts, (3) drivers of permafrost change, (4) hydrologic cycle change and its consequences, (5) flora and fauna responses to environmental change, and (6) the biogeochemical mechanisms driving change in soil carbon pools. ABoVE will emphasize integration and synthesis across these science themes.

Another relevant activity is the European Space Agency (ESA)-funded Data User Element (DUE) Permafrost Project (2009-2012). Its objective is to establish a permafrost-related monitoring system based on satellite remote sensing data. The international permafrost research community requires permafrost-related products at a variety of spatial scales. To this end, user organizations representing permafrost field investigators and modelers were involved in the early stages of the DUE Permafrost Project to define satellite-derived products and a satellite-based obser-

[1] See http://csc.alaska.edu/projects/integrated-ecosystem-model.

vation strategy. The products identified by users were regional- and circumpolar-scale products consisting of land surface temperature (LST), surface soil moisture (SSM), frozen/unfrozen states of the ground surface, terrain parameters, land cover, and surface waters. The suite of data products developed and evaluated during the course of DUE Permafrost can be visualized and accessed via a Web-GIS Service[2] and downloaded along with product documentation from PANGAEA (DUE Permafrost Project Consortium, 2012).

Experimental applications of the DUE Permafrost products are currently being developed. For example, within the European Union (EU)-funded project *Changing Permafrost in the Arctic and its Global Effects in the 21st Century* (PAGE21), which started in 2011, modeling groups are making extensive use of data products from DUE Permafrost. Experiments include the integration of satellite-derived products into permafrost models and the evaluation of output from regional climate models (e.g., spatial patterns of SSM and LST).

Another relevant activity is the Department of Energy-sponsored NGEE (Next Generation Ecosystem Experiments),[3] which seeks to quantify the physical, chemical, and biological behavior of terrestrial ecosystems in Alaska through a coordinated set of investigations for improved process understanding and model representation of ecosystem-climate feedbacks. NGEE's initial research will focus on thaw lakes, drained thaw lake basins, and ice-rich polygonal ground on the North Slope (Barrow, Alaska). The goal is to produce a process-based ecosystem model that can demonstrate the evolution of Arctic ecosystems in a changing climate with a high-resolution Earth System Model grid cell. NGEE will also include mechanistic studies in both the field and the laboratory; modeling of critical and interrelated water, nitrogen, carbon, and energy dynamics; and characterization of important interactions from molecular to landscape scales that drive feedbacks to the climate system.

A fourth related activity is the Integrated Ecosystem Model (IEM)[4] for Alaska and Northwest Canada which is an effort to try and forecast landscape change within the Alaska and Northwest Canada region. This 5-year project, which started in 2011, uses three ecosystem models that link changing climate scenarios to different ecological processes. The goal of IEM is to generate maps and other products to illustrate how Arctic and boreal landscapes are expected to change due to climate-driven changes to vegetation, disturbance, hydrology, and permafrost. The products will also provide the uncertainty in the expected outcomes. IEM is sponsored by the Arctic Landscape Conservation Cooperative, the Northwest Boreal Landscape Conservation Cooperative, the U.S. Geological Survey Alaska Climate Science Center, and the Western Alaska Landscape Conservation Cooperative and is composed of members from different research communities, including the Geophysical Institute Permafrost Lab, the Institute of Arctic Biology, and the Scenarios Network for Alaska and Arctic Planning.

WORKSHOP DESCRIPTION

To complement the activities above, NASA asked the NRC to organize a workshop to explore opportunities for using remote sensing to advance our understanding of permafrost status and trends and the impacts of permafrost change (see Appendix C for Statement of Task). The workshop brought together experts from the remote sensing community with permafrost and ecosystem scientists. Participants represented academia, federal agencies, national laboratories, and the private sector; there was also international participation. In planning the workshop, the committee considered the other past and ongoing activities (discussed in the previous section) in developing the workshop agenda. The workshop discussions were designed to help the community articulate gaps in current understanding and identify potential opportunities to harness remote sensing techniques to better understand permafrost, permafrost change, and the implication of this change for the global carbon cycle in permafrost areas (see Appendix A for abstracts of the presentations and Appendix B for the workshop agenda and participant list).

Participants at the workshop addressed questions such as

[2] See http://www.ipf.tuwien.ac.at/permafrost/.

[3] See http://ngee-arctic.ornl.gov/about.

[4] See http://csc.alaska.edu/projects/integrated-ecosystem-model.

- How might remote sensing be used in innovative ways?
- How might remote sensing enhance our ability to document long-term trends?
- Is it possible to integrate remote sensing products with ground-based observations and assimilate them into advanced Arctic system models?
- What are the expectations of the quality and spatial and temporal resolution possible with such approaches?
- What prototype sensors (e.g., the airborne UAV SAR, AirMOSS, AIRSWOT, MABEL, IceBridge) are available and might be used for detailed permafrost studies after ground calibration to address many scientific and practical questions related to changes in high-latitude permafrost, including carbon cycle studies?

The workshop was divided into three plenary sessions: (1) measurement of permafrost properties, (2) measurement of related ecological characteristics, and (3) emerging remote sensing technologies and approaches for studying permafrost. Significant time was also spent in breakout groups to discuss current and future approaches for remotely sensing permafrost and ecological variables (see Tables 2.1 and 2.2). The focus of the workshop was on Arctic permafrost, although many of the remote sensing technologies discussed in the report could be applied in general to other types of permafrost such as mountain permafrost.

STRUCTURE OF THE REPORT

This report summarizes the views expressed by individual workshop participants. Although the committee is responsible for the overall quality and accuracy of the report as a record of what transpired at the workshop, the views contained in the report are not necessarily those of all workshop participants, the committee, or the NRC.

Chapter 1 (this chapter) provides background on permafrost characteristics, how the permafrost changes over time, and how it may impact other ecological components. It also describes the role of remote sensing to map and monitor changes in Arctic permafrost. Related activities dedicated to better understanding permafrost through remote sensing are also discussed. Chapter 2 describes various permafrost properties and permafrost-related ecological characteristics that can be remotely sensed to understand and detect changes in permafrost. Chapter 3 looks to the future and discusses systems (airborne and satellite) that can be applied immediately, in the near term, and in the long term to study permafrost.

2

Remote Sensing Technologies to Directly and Indirectly Measure Permafrost and Permafrost-Related Ecological Characteristics

Satellite, airborne, and surface-based sensors operating from the very-low-frequency (VLF) range to the optical region of the electromagnetic (EM) spectrum have been used for remote sensing of permafrost landscapes for several decades (Duguay et al., 2005; Hall, 1982; Kääb, 2008; Westermann et al., In press). As discussed in the previous chapter, permafrost can be studied by remote sensing directly and indirectly through the measurement of permafrost-related ecological variables. Measurements of subsurface permafrost could be based indirectly on remotely sensed observations of the climate and hydrological controls (such as air and surface temperatures, snow depth, and soil moisture) and on observations of the surface expressions of subsurface phenomena or on measurements of subsurface properties directly. The former observations are, for example, obtained using fine-resolution spaceborne and airborne optical systems, which have been used for mapping ground cover, topography, and changes in permafrost. Satellite and airborne optical and IR (infrared) sensors have been used to determine surface topography, soil properties, and surface temperature and to map thermokarst (Duguay et al., 2005; Kääb, 2008).

Active microwave sensors have also been used for permafrost monitoring and mapping. Ground-penetrating radars have been used to map the thickness of both the active layer, particularly during the winter season before the onset of surface melt, and buried ice in permafrost during both the winter and summer (Arcone et al., 2002; Hubbard et al., 2013; Leuschen et al., 2003; Westermann et al., 2010; Yoshikawa et al., 2006). Radar backscatter from a distributed target depends on surface roughness, dielectric properties, and the internal structure of the subsurface. In addition, the backscattered signal is strongly modulated by the presence of water and has been exploited for mapping freeze/thaw conditions, surface soil moisture and soil moisture profiles under favorable conditions, and surface topography (Alasset et al., 2008; Bartsch et al., 2012; Boehnke and Wismann, 1996; Komarov et al., 2002; Sabel et al., 2012; Strozzi et al., 2010; Watanabe et al., 2012). InSAR has been utilized to study active layer changes and long-term surface deformation (Liu et al., 2010; Singhroy et al., 2007).

Coarse-resolution passive microwave sensors (with tens of kilometers footprints) are used to determine soil moisture and snow conditions (Yubao et al., 2010). Both spaceborne and airborne LiDAR (LIght Detection And Ranging) have been used for fine-resolution mapping of surface topography of regions underlain by permafrost (Hubbard et al., 2013; Jones et al., 2013; Figure 2.1).

During the workshop, two tables were developed to better structure the discussion on permafrost research needs and the relevant remote sensing methods and sensors that may address these needs (Tables 2.1 and 2.2).[1] Table 2.1 consists of permafrost processes and properties, and Table 2.2 consists of ecological variables that are categorized by six permafrost-related research ecological categories: climate, topography, geology and soil, hydrology, vegetation and land cover, and

[1] These two tables are not comprehensive; rather they serve as a tool for organizing the information that was discussed at the workshop.

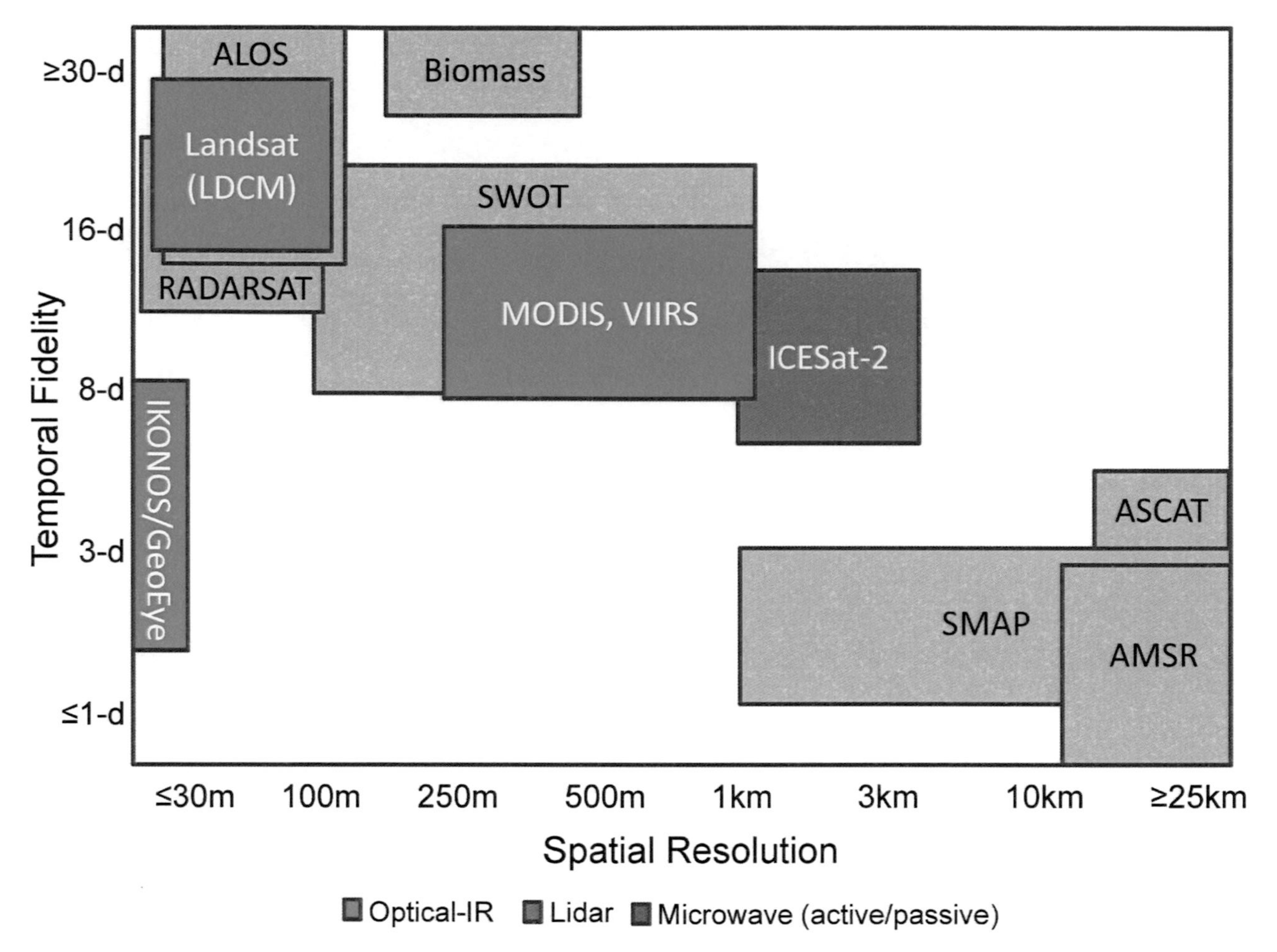

FIGURE 2.1 Space-time plot of selected near-term (2013-2020) satellite sensor observations with potential relevance for permafrost. ALOS-2 (L-band SAR) has a planned launch in 2014. ESA Biomass Earth Explorer mission will be the first satellite P-band SAR with a potential launch in 2019. Landsat 8 was launched in February 2013 and utilizes two sensors, the Operational Land Imager and the Thermal InfraRed Sensor. RADARSAT (C-band radar data) refers to the RADARSAT Constellation Mission, which is scheduled for launch in 2018. IKONOS and GeoEye/RapidEye are commercial optical-NIR (near infrared). SWOT (Surface Water Ocean Topography) is a Ka-band SAR altimeter and radar interferometer with a potential launch in 2020. VIIRS, a subsequent mission to MODIS, is a scanning radiometer that collects visible and infrared imagery and radiometric measurements. ICESat-2 (LiDAR) is a second-generation orbiting laser altimeter scheduled for launch in 2016. ASCAT (Advanced SCATterometer) is a C-band radar that was launched in 2006. SMAP (L-band SAR) is scheduled to launch in late 2014. AMSR (Advanced Microwave Scanning Radiometer) is a passive microwave radiometer and does not currently produce data. HyspIRI (Hyperspectral Infrared Imager; not shown in figure), which will include a combination of VIS-IR and TIR instruments, is still in the study stage. This figure does not include all upcoming missions applicable to permafrost remote sensing studies (e.g., ESA Sentinels 1 and 2 [SAR and Landsat-type optical sensors]; gravimetry missions; X-band SAR, TerraSAR-X, and Tandem-X). SOURCE: Image courtesy of John Kimball, University of Montana.

greenhouse gases. Each variable in both tables has a description of any relevant ecological indicators (second column); remote sensing techniques that currently exist to measure these variables (the third column) or their ecological indicators (the fourth column); remote sensing technologies that may become available in the future (the fifth column); the available (the sixth column) and desirable (the seventh column) spatial and temporal resolution of the relevant remote sensing products; and examples from the published literature (last column). The following discussion is structured in accordance with the categories presented in Tables 2.1 and 2.2.

Of the more than 60 permafrost and related ecological variables that were discussed during the workshop, those listed in Box 2.1 emerged as having the most impact in advancing the current state of knowledge of permafrost landscapes, if they could

BOX 2.1
Important Permafrost and Related Ecological Variables to Measure with Remote Sensing

Active layer thickness
Ground ice (volume and morphology)
Snow characteristics (extent, water equivalent, depth, density, conductivity)
Surface topography (static, macro-, and micro-)
Longer-term surface subsidence
Thermokarst distribution
Surface water bodies (including dynamics, redistribution)
Surficial geology-terrain units (including lithology, bedrock)
Soil organic layer (thickness, moisture, conductivity)
Land cover (including spectral vegetation indices)
Vegetation structure and composition
Methane (flux or concentration)
Water vapor flux
Carbon dioxide (flux or concentration)
Land surface (skin) temperature
Subsurface soil temperature
Seasonal heave/subsidence
Soil moisture
Biomass (above ground)

be measured through remote sensing. It is important to note that it may not be currently possible or even feasible to remotely sense some of these properties in the near future.

PERMAFROST PROPERTIES AND PROCESSES

Permafrost

Active layer thickness is defined as the depth of a near-surface layer of soil or rock that experiences periods of freezing and thawing during an annual cycle. The freezing/thawing processes may produce some distinct geomorphological features such as sorted and unsorted circles and other small-scale patterned ground (Figure 2.2). The term "active layer" is most commonly used to describe a seasonally thawed layer above permafrost that thaws through during summer and completely refreezes during the next winter. The maximum depth of the summer thaw is usually observed at the very end of the warm period (August, September, or even October in the Northern Hemisphere, depending on site location) and may vary between a few decimeters to several meters. This maximum depth of thaw is called the "active layer thickness" (ALT). Another variable related to seasonal freezing and thawing is called the "depth of seasonal freezing." This variable may be observed and measured in the area where permafrost is either completely absent or absent in at least the few upper meters. The seasonally frozen layer forms during the cold part of the year and completely thaws during the summer. Both of these variables are relevant to permafrost. Increase in the maximum depth of seasonally thawed layer (or in the ALT) indicates the beginning of permafrost degradation and may lead to local permafrost instability if any substantial amount of ground ice is present in the near-surface permafrost. Inversely, if the increase in the seasonally frozen layer thickness exceeds the maximum thickness of the seasonally thawed layer, then it may lead to the formation of new permafrost. Both of these variables are important for functioning of the ecosystem and for the stability of infrastructure in cold regions.

Several traditional methods reviewed by Hinkel and Nelson (2003) are used to determine the inter-annual and long-term changes in thickness of the active layer: mechanical probing once annually, frost (or thaw) tubes, and interpolation of soil temperatures obtained by data loggers. The ability to use remote sensing methods to directly estimate the depth of the active layer over large synoptic areas is limited at this moment, but a few promising techniques based on lower-frequency microwaves have been proposed recently. For small areas of coverage, surface-based ground penetrating radar (GPR) systems have been proven to provide high-quality information about the ALT (Hubbard et al., 2013). These sensors, however, cannot feasibly be used to produce such information over large regions. Low-frequency (i.e., long wavelengths) airborne or spaceborne synthetic aperture radars (SARs) are able to cover large areas rapidly. They have the capability to penetrate the ground to varying degrees, depending on the liquid water content, layering structure, and texture of soils. Although specific demonstration of ALT retrieval using SAR systems over permafrost soil has not yet taken place, it has been shown that retrieval of soil moisture for a range

FIGURE 2.2 Active layer features. The left shows "sorted ground," and the right shows "small-scale patterned ground." SOURCE: Photos by Guido Grosse.

of textures and temperatures, as well as for a variety of land cover types, is possible with good fidelity (Tabatabaeenejad et al., in review). Recent observations using the AirMOSS and the UAVSAR airborne synthetic aperture radars (Tabatabaeenejad and Moghaddam, 2011) have shown the potential for this capability. Additionally, the airborne electromagnetic (AEM) method[2] was recently proposed as a way to estimate the ALT (Pastick et al., 2013).

Liu et al. (2012) recently proposed an indirect method of estimating the dynamics of the ALT using seasonal ground-surface subsidence trends derived from InSAR (interferometric synthetic aperture radar; Figure 2.3). This method requires knowledge of many other characteristics of the active layer, such as porosity and water content. These requirements may limit the ability to use this method. However, other remote sensing methods may be used to estimate the necessary variables and help to apply this method to the areas where the required information is currently unavailable. Another approach in estimating the ALT may be the application of physically based permafrost dynamics models (Riseborough et al., 2008; Sazonova and Romanovsky, 2003; Sazonova et al., 2004). Again, the required input data for these models may be derived from existing remote sensing products.

[2]AEM relies on the physics of electromagnetic induction to detect physical properties from the near surface down to several hundred meters below ground. Inductive electromagnetic methods are primarily sensitive to the electrical characteristics of geological materials that, in turn, are a function of properties such as unfrozen water content, lithology, and salinity. For more information on AEM, see Burke Minsley's abstract in Appendix A.

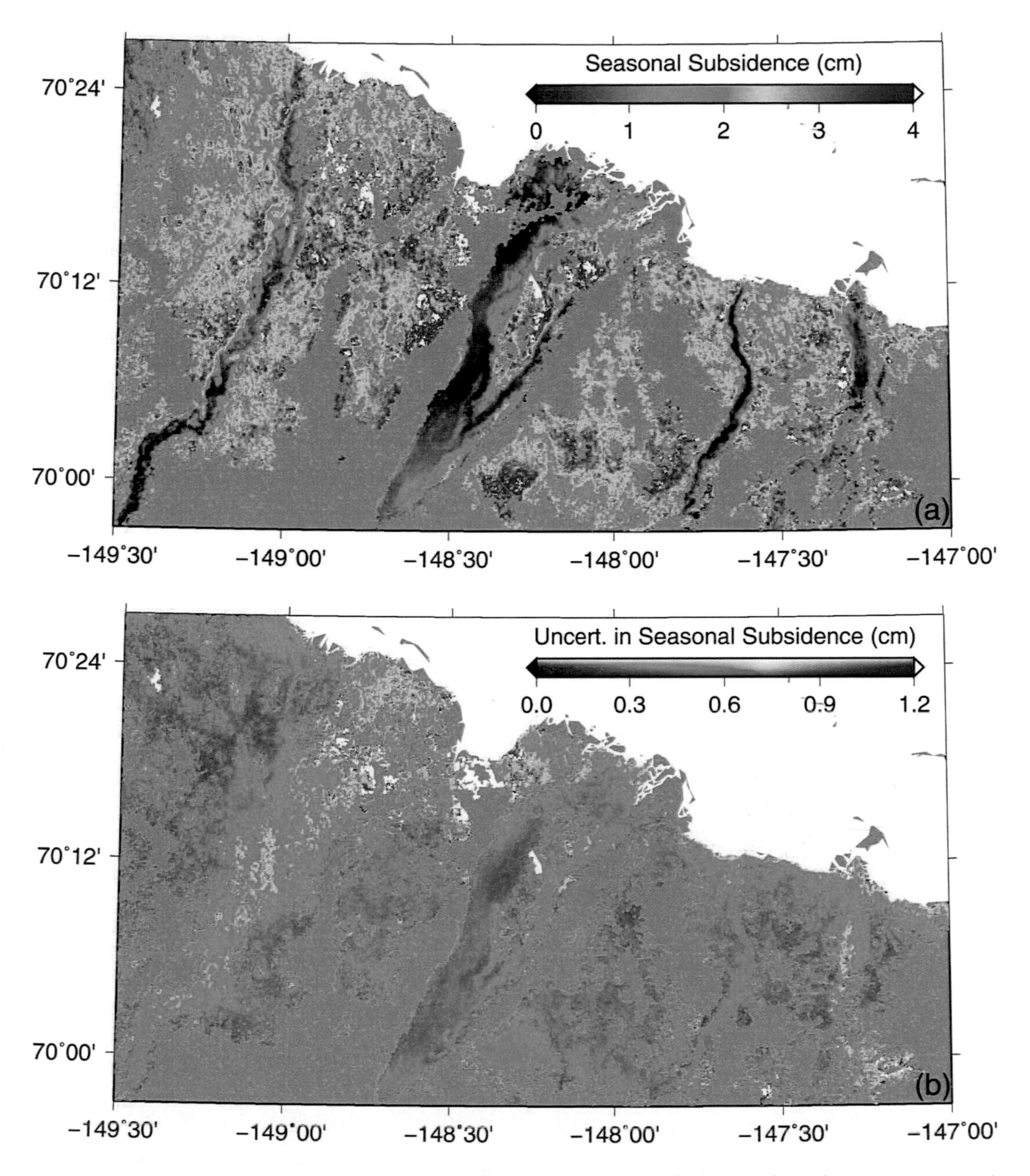

FIGURE 2.3 Mapping the 1992-2000 average ALT in Prudhoe Bay using InSAR methods. SAR phase observations can provide measurements of surface dynamics, which can be used to indicate surface change and measure cm-scale surface deformation. An algorithm was developed to estimate long-term average ALT using thaw-season surface subsidence derived from spaceborne InSAR measurements. The algorithm uses a model of vertical distribution of water content within the active layer accounting for soil texture, organic matter, and moisture. The estimated ALT values match in situ measurements at Circumpolar Active Layer Monitoring (CALM) sites within uncertainties. SOURCE: Liu et al., 2012.

Many workshop participants noted that a desirable resolution for the ALT is 5 cm vertical and 30 m horizontal. The minimum temporal resolution of measurements would be once per year at the end of the warm season (mid-August to late September depending on latitude). However, it is desirable to have several measurements per warm season to establish the rate of thaw depth propagation.

Ground ice (volume and morphology). The presence of ground ice makes permafrost a unique component in the geological system (Figure 2.4). Several permafrost properties depend on the amount of ice included and the specific geometric forms of these inclusions. If ice is present, then all properties of permafrost become temperature dependent with a major threshold at the melting point of ice. Because at some locations the volume of ice in permafrost may exceed 80 percent or even 90 percent of the total volume, dramatic changes in the environment could be expected upon thawing. Permafrost can also be very ice poor, and the volumetric ice content may not exceed a few percent in other locations. In this situation, thawing of permafrost will not produce any significant changes in micro-topography. However, even in this situation, the loss of pore ice can affect the physical and biogeochemical properties of the subsurface substrate.

The amount of ground ice is measured in the field by permafrost coring and sampling of the natural permafrost exposures (French and Shur, 2010; Kanevskiy et al., 2013; Murton, 2013). The site-specific data on ground ice obtained in the field may be extrapolated

FIGURE 2.4 Large syngenetic ice wedges that developed in eolian silt during the late Pleistocene exposed in a ~25-m-high bluff along the Itkillik River in northern Alaska (see Kanevskiy et al., 2011; photo by T. Jorgenson). Syngenetic permafrost is formed when soil freezes and becomes permafrost as the soils are being deposited by wind, water, and/or gravity.

to the larger region using the knowledge of geological structure and the permafrost history of the entire area of investigation. Remote sensing products could be efficiently used for this extrapolation. The remote sensing methods that could be used to directly estimate the relative amount of ground ice in permafrost include AEM; the application of AEM (Minsley et al., 2012) for permafrost measurements, including ALT and ground ice content, has been demonstrated. A few studies with GPR were conducted to map ground ice with successful results under favorable conditions (Arcone et al., 2002; Leuschen et al., 2003; Yoshikawa et al., 2006). Most of these measurements were with surface-based systems. Airborne GPR with SAR and tomographic processing have the potential to map ground ice over large areas. In addition, ultra-wideband SARs operating in the high-frequency (HF) and very-high-frequency (VHF) regions might be useful for ground ice mapping. Promising indirect methods may be based on the estimation of total subsidence in the ground surface produced by ice masses melting away. The difference in topography between intact surfaces and the bottom of the thermokarst depressions measured using remote sensing techniques may be used to estimate the volume of melted ice. High-resolution optical sensors and airborne LiDAR may be used for this purpose.

Several workshop participants indicated that a horizontal resolution of 1 to 10 m is sufficient for this permafrost characteristic. They noted that the accuracy of ice volume estimations of 10 percent or more of the total volume would be better.

Subsurface soil temperature. According to its definition, the presence or absence of permafrost depends on the temperature of the soil. Many of the properties of permafrost, including those important to engineering, also depend on temperature. Soil temperatures are obtained in the field by lowering a calibrated temperature sensor into a borehole or recording temperature from multisensor cables permanently or temporarily installed in the borehole. Measurements may be recorded manually with a portable temperature logging system or by data loggers. No remote sensing techniques will allow direct measurement of subsurface soil temperature. However, the remote sensing derived products of land surface (or "skin") temperature (LST),[3] snow water equivalent, and snow depth (see the Climate section) may be used in physically based permafrost thermodynamic models to indirectly calculate subsurface soil temperature. Microwave radiometers operating in the ultrahigh-frequency (300-1,000 MHz) range of the EM spectrum are being explored for measuring subsurface temperature (Matti Vaaja and Hallikainen, 2013). Workshop participants noted that when using this modeling approach for estimating subsurface temperatures a horizontal resolution of LST data on the order of 30 m would be optimal.

Permafrost presence/absence. Maps of permafrost distribution are needed for infrastructure planning and various types of land management. At the same time, permafrost distribution maps are presently available only in generalized form at global to regional scales. Fine-resolution maps have not yet been developed for most permafrost-affected regions. Recent studies show good potential of AEM to directly map near-surface permafrost. Some workshop participants noted that this task is not trivial and requires additional information on geological structure and other environmental characteristics of the mapped area. Improvements here could lead to the development of fine-resolution maps of permafrost over large areas. Several workshop participants noted that the required horizontal resolution depends on the scale of investigation. For circumpolar mapping, a resolution of several hundred meters would be sufficient; a horizontal resolution of 1-10 m is needed for local maps of permafrost distribution.

Depth to top of permafrost. This permafrost property is critical for estimating the long-term rate of permafrost thaw. "Talik"—which occurs when the ground layer above permafrost does not completely freeze, even during the winter months—plays an important role in surface and subsurface hydrology and in the permafrost-carbon cycle-climate feedbacks. The thermal conditions in this layer are suitable for supporting year-round decomposition of the organic matter previously sequestered in the near-surface permafrost. Depth to top of permafrost is also important for engineering and infrastructure development on degrading

[3] Land surface temperature, which is different from air temperature, is the temperature of the uppermost surface of objects on the Earth surface as detected by remote sensors.

permafrost. The AEM method can be used to estimate this variable. There are practically no indirect methods that can be used to address this question, except for the application of physically based permafrost thermodynamics models. As discussed by many participants, the required horizontal resolution depends on the expected use of the product. For circumpolar mapping, a resolution of several hundred meters would be sufficient. For developing local maps of permafrost distribution, it is desirable to have 1 to 10 m horizontal resolution.

Permafrost thickness and 3D geometry. Knowledge of permafrost thickness and 3D geometry is important in relation to understanding permafrost extent and surface and subsurface hydrology, including thermokarst lake dynamics. Development of new thermokarst lakes and an extension of existing lakes lead to changes in permafrost geometry below lakes and near lake margins. The related talik formation and extension is an important process leading to the production of a significant amount of methane, a critically important greenhouse gas (Walter et al., 2006, 2007a, b). AEM has been used to estimate these permafrost properties (Jepsen et al., 2013; Minsley et al., 2012). SARs operating in the VHF and P-band part of the spectrum with tomographic capability have the potential to provide information on permafrost thickness and 3D geometry. These include GeoSAR, AirMoss, and unmanned aerial vehicle (UAV) SARs. The radar depth sounder/imager currently being operated on NASA IceBridge might also be useful for this purpose (Rodriguez-Morales et al., 2013).

In particular, ultra-wideband radars for 3D imaging of the ice-bed interface of polar ice sheets have been developed over the past few years, applying basic concepts of tomography using a sequence of 2D image slices, much like what is done with X-ray tomography. Many technical limitations in this technology have recently been overcome through advances in RF and microwave, as well as digital technologies spawned by the communication industry. Radars with multiple receivers and transmitters, large bandwidths, and advanced SAR processing algorithms have been developed and used for fine-resolution 3D measurements over ice sheets (Jezek et al., 2011; Paden et al., 2010; Rodriguez-Morales et al., 2013). Such radars can potentially be used to measure the ALT and map ground ice.

Near-surface seasonal freeze/thaw state. Although not unique to permafrost landscapes, near-surface freeze/thaw state and its duration, timing, and spatial characteristics may have strong correlation with presence/absence and physical properties of permafrost. Active microwave sensors have been used extensively for detecting seasonal freeze/thaw (e.g., Kimball et al., 2006; McDonald et al., 2004). This measurement is among the most accurate and sensitive measurements possible with microwave sensors, because the backscattering properties of dielectric material that include water (such as soils and vegetation) are strongly altered when transitioning between frozen and thawed states. In frozen form, soils and vegetation have very low backscattering strengths. With increasing water content (e.g., transition of ice into liquid water), the dielectric constant of soils and vegetation components increases rapidly and generally causes a marked increase in backscattering across sections measured by the active sensors. Using time-series analysis of microwave measurements, it is possible to identify clear thresholds for freeze/thaw transitions for various landscapes. The thresholds can vary significantly depending on whether the ground is covered by vegetation and, if so, by the type and density of vegetation. Furthermore, freeze/thaw state has environmental indicators that may also be characterized with other remote sensing techniques. These include presence or absence of liquid water, surface heave or subsidence, overlying vegetation characteristics, and snow properties. A variety of existing and planned sensors are suitable candidates for retrieving these properties, including repeat-pass C-band or L-band InSAR (for surface subsidence), LiDAR (for surface subsidence, vegetation properties, snow), and various SAR instruments (for presence of liquid surface water).

PERMAFROST-RELATED ECOLOGICAL VARIABLES

Climate

Climate is, of course, a critical driver for the evolution of frozen ground. On the landscape scale, mean climate conditions have led to the past creation of permafrost and ground ice; recently they have also led to its degradation. On the plot scale, microclimate features such as topographic-related variations in soil moisture and temperature are the results (and further drivers) of the forms of change most closely associated with permafrost, such as patterned ground and thermokarst. When in equilibrium, LST is the result of a balance of upward and downward radiative sensible and latent heat fluxes and is impacted by the vegetation, albedo, and underlying soil properties, such as thermal conductivity.

Land surface skin temperature and atmospheric temperature. Profiles of these two variables are both accessible from a number of historic and current satellites, including sensors on NOAA's GOES and POES platforms, AIRS on Aqua, and CrIS on SNPP. LST is also available from AMSR, SSM/I, MODIS, SNPP, AVHRR, Landsat, ASTER, and airborne sensors using either infrared or microwave technologies (Hachem et al., 2009; Jones et al., 2010; Kimball et al., 2009). Land-based in situ air temperature measurements are typically made between 1 and 3 m above ground and will differ somewhat from LST (Hachem et al., 2012).

Precipitation. As a source of moisture, precipitation is also an important driver for the evolution of permafrost and ground ice. Quantitative precipitation estimates (QPE is "how much" precipitation as opposed to indicators of whether or not it is precipitating) are generated from operational meteorological satellites, such as the GOES and POES platforms in the United States. QPE algorithms are designed to work best in the contiguous United States and are not optimal for solid or mixed precipitation or for high latitudes. However, this information is challenging to verify because of the extreme difficulty of accurately measuring solid precipitation, even near the ground. Participants reported that the launch of the Global Precipitation Mission may show improvements in satellite estimates of precipitation in cold regions; airborne testing with sensor prototypes was promising.[4]

Snow on the ground. A critically important driver for the evolution of ground temperature is snow on the ground because of its powerful insulating properties. Key properties of snow for permafrost interactions include its depth, density, snow water equivalent (SWE), and thermal conductivity. Sensors with optical bands, such as MODIS, are relatively successful at mapping the snow-covered area (SCA) at the 500-m pixel scale, particularly in treeless areas (Hall and Riggs, 2007). Other satellite-borne sensors, such as passive microwave (e.g., AMSR, SSM/I), have been used to develop SWE products with mixed results (Derksen et al., 2003; Hancock et al., 2013). The global and region-specific algorithms developed to date vary considerably in their uncertainties. Radar-based approaches have been proposed to more accurately measure SWE and have been flown successfully on aircraft (Rott et al., 2010, 2012; Xu et al., 2012). The ultra-wideband radars (i.e., 2-8 GHz and 12-18 GHz) being flown as a part of Operation IceBridge to measure thickness of snow over sea ice also have the potential to measure the thickness of snow over land, as discussed at the workshop.

Snow depth on the ground is highly variable on scales from 10 cm and larger, in part because the pack evolves differently when it interacts with vegetation, and wind may redistribute snow on the ground, packing it into topographic and biological crevices. Modeling can estimate snow depth from snow-covered areas, but only during the depletion season, and it is limited by the large pixel size of the SCA products (such as 500 m from MODIS). Many workshop participants noted that this scale is sufficient for large landscape-scale estimates of snow cover, but not for studying snow interaction with individual thermokarst features or ice-wedge polygons (1-10 m scale). Currently, this can only be done from aircraft or commercial/defense satellites.

For large-scale permafrost modeling, researchers generally tend to use meteorological inputs from atmospheric reanalyses (e.g., Anisimov et al., 2007; Arzhanov et al., 2008; Mugford and Dowdeswell, 2010) rather than from remote sensing directly. Two notable exceptions are Marchenko et al. (2009) and Langer et al. (2013), who integrated satellite-derived LST and SWE in permafrost model experiments. Atmospheric reanalyses are historical runs of atmospheric or coupled oceanic-atmospheric models that assimilate remote sensing and in situ (especially radiosonde) observations. Analyses products are convenient because output is gridded and different state variables

[4] http://pmm.nasa.gov/GCPEx.

(e.g., air temperature, precipitation, wind) are forced to have some physical consistency and operate on first principles. While the spatial resolution of reanalysis products is continually improving, they are currently available at the "large landscape scale," for example, 0.5 degree latitude and 0.7 degree longitude for NASA's Modern Era-Retrospective Analysis for Research and Applications (MERRA) in 2012 and approximately 0.3-degree (32-km) grid resolution for the North American Regional Reanalysis (NARR) product. Another advantage of reanalysis products is that many are now available for multiple decades such that interannual and decadal-scale variability is encompassed, including variability attributed to large-scale modes like El Niño-Southern Oscillation, the Pacific Decadal Oscillation, and the Arctic Oscillation. Shorter-term observations from airborne or satellite missions may not capture the full range of variability evident in longer climate records.

For researchers interested in the evolution of plot- and watershed-scale features on the landscape, the spatial resolution of current satellite remote sensing and reanalyses products is sufficient for some variables (e.g., air temperature) and insufficient for others (e.g., LST, precipitation, snow on the ground). Many workshop participants noted that airborne remote sensing observations are particularly useful for fine-spatial-resolution measurements of LST and mapping snow on the ground, but are difficult to support for routine temporal repetition of measurements.

Topography

Surface topography. Workshop participants noted that surface topography is an important variable for permafrost landscapes. In particular, it is important to measure thermokarst distribution by identifying elevation changes in areas where active thermokarst is occurring and by estimating loss of ice content in these thermokarst areas (Jones et al., 2012; Short et al., 2011). Topographic variables also provide essential information on the presence/absence of permafrost via "integrated terrain units" that combine slope, aspect, and elevation along with other indirect indicators. Topographic variables (such as longer-term surface subsidence and seasonal heave and subsidence) are also useful for estimating the topographically influenced distribution and variability of incident radiation across the surface, which can be used for traditional mapping approaches as well as for input in some permafrost models that simulate radiation balance. Because topography can be directly measured, many workshop participants believe that there is no pressing need to measure ecological indicators of topography, although a number of other variables (such as lake extent and vegetation cover change) are a result of and thus indicators of topographic change. Indirect indicators do have the advantage of being more readily available, for example, from stereo-interpretation of high-resolution optical remote sensing images, although they may not be as accurate as more direct remote sensing approaches.

Currently, the best system to measure topography and topographic change is airborne LiDAR, which provides relatively simple and direct yet highly accurate data on elevation, from which surfaces can be gridded at various spatial resolutions depending on the density of LiDAR sampling (e.g., Jones et al., 2011). LiDAR has the added advantage of providing structure information on vegetation canopies along with the underlying topography. InSAR can also be used to measure topography (Figure 2.5; Chen et al., 2013; Liu et al., 2012; Short et al., 2011). InSAR uses phase information of reflected radiation, which is obtained by two displaced antennas on two or more aircraft or satellites passing over the same area to determine surface topography. The horizontal resolution of spaceborne InSAR digital elevation models (DEMs) ranges from about 30 to 100 m and vertical resolution from 10 to 15 m. For airborne InSAR DEMs, the horizontal resolution ranged from about 1 to 10 m and vertical resolution from 50 cm to 2 m. Several aircraft LiDAR and SAR instruments are suitable for topography mapping, including those from commercial LiDAR providers. There is currently no space-based LiDAR mission available for topographic mapping, but the Tandem-X InSAR mission is currently active and additional coverage should allow repeat DEM differentiation through time. As some participants noted, SAR sensors onboard previously active platforms (e.g., ALOS, ENVISAT, RADARSAT), could also be used for InSAR-based topographic change mapping (subject to various temporal decorrelation limitations). Data from IceSat-I can also be used for topographic mapping, although the large footprint

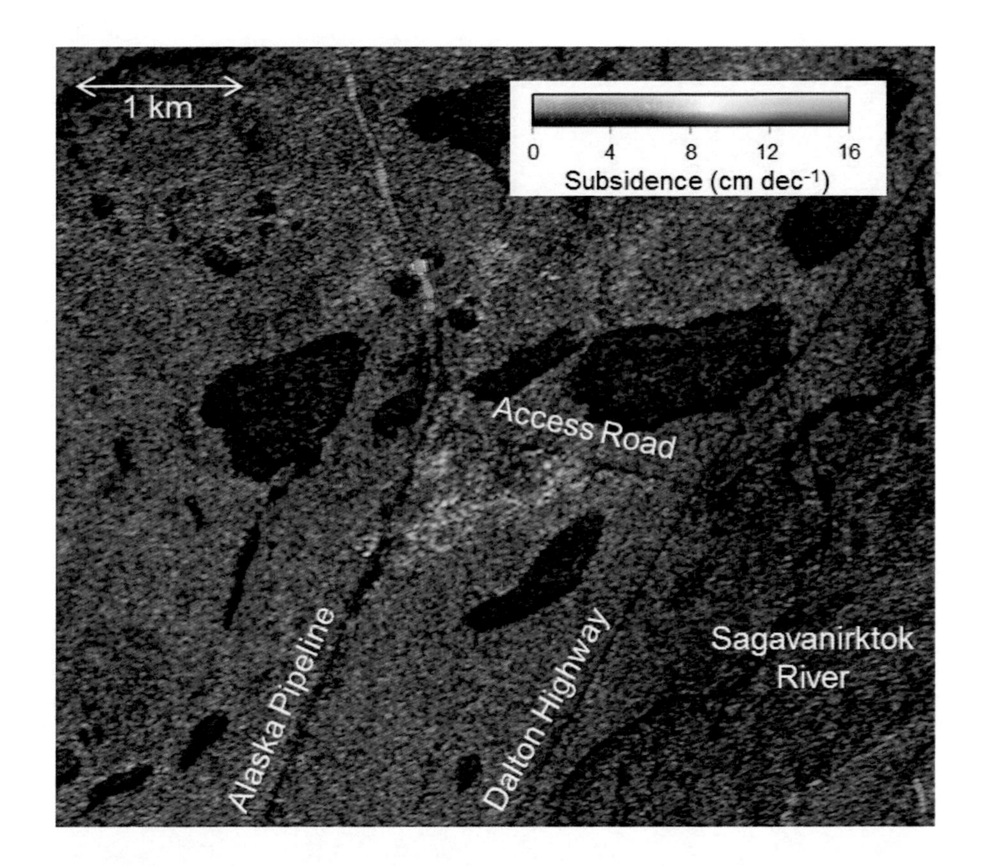

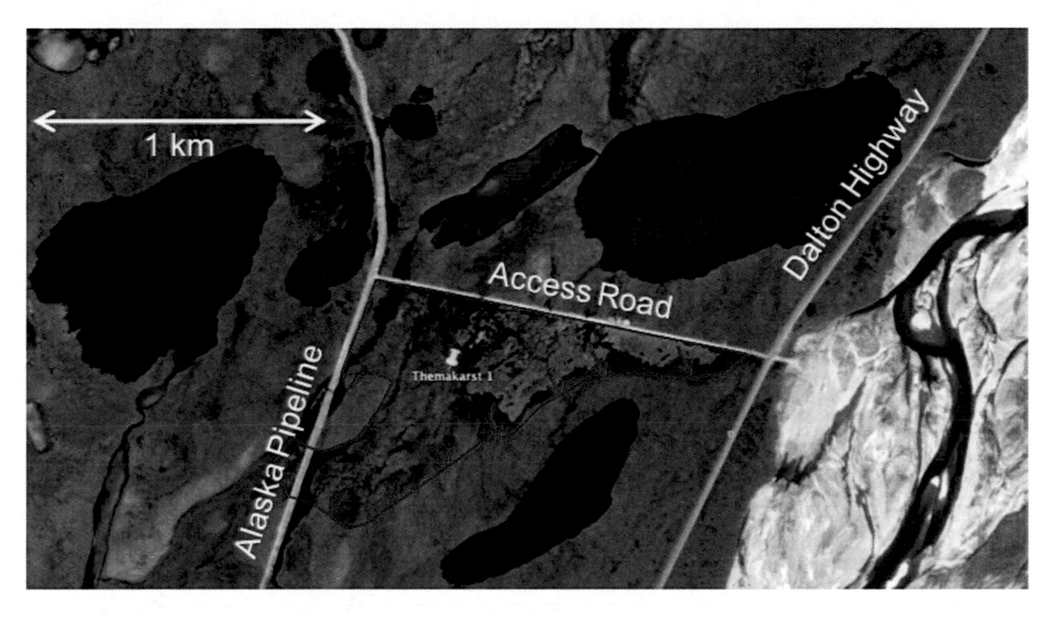

FIGURE 2.5 Thermokarst dynamics using InSAR. Anomaly in ALOS (Advanced Land Observing Satellite) subsidence trends near Deadhorse, Alaska (top). Google Earth image of thermokarst feature in Deadhorse, Alaska (bottom). Presented by Kevin Schaefer.

and wide spacing makes it of limited utility for many of the topographic changes taking place in the Arctic (e.g., thermokarst features).

Airborne and spaceborne photogrammetery is also being used to generate topographic maps using a computer vision technique called Structure from Motion and a new class of software. Some participants noted that the use of this technique is growing quickly because the cost of acquisition is far less than airborne LiDAR, and, combined with ground control, errors are comparable to those from LiDAR or smaller. There also was mention of change detection software developed by both military and commercial interests that could facilitate change mapping of thermokarst features using various data sources (aircraft or satellite acquired).

The planned future sensors that could prove to be most useful for topographic mapping include the U.S. L-band InSAR mission (currently scheduled for a 2021 launch date) and the ICESat-2 ATLAS photon counting LiDAR mission (currently scheduled for a 2016 launch).

For developing regional maps of topography, many participants indicated that it would be desirable to have 1 to 10 m horizontal resolution; the higher resolution would be more applicable to local-scale studies of thermokarst features. At more global scales, topographic horizontal resolution would be beneficial at 30 m or less with a vertical accuracy less than 10 cm.

Geology and Soil

Surficial geology, as it relates to differentiating unconsolidated deposits with varying soil/sediment textures, depositional processes, and landscape age (Jorgenson et al., 2008; Kanevskiy et al., 2013; Kreig and Reger, 1982), is critical to assessing permafrost distribution and ground ice characteristics, said several workshop participants. Soil texture affects the moisture-holding capacity and development of segregated ice (Kreig and Reger, 1982), depositional history affects patterns of syngenetic and epigenetic[5] permafrost formation (Shur and Jorgenson, 2007), and age affects the amount of time over which ice can aggrade or transition through complicated permafrost histories (French and Shur, 2010).

Surficial geology. Surficial geology and the related concepts of terrain units (Kreig, 1977), engineering geology (Carter and Galloway, 1985), landform soils (Kreig and Reger, 1982), and geomorphic units (Jorgenson et al., 1998) have traditionally been mapped through photo-interpretation of aerial photography, especially with stereo-pairs that enhance 3D recognition of terrain characteristics. The combination of photo-interpretation and borehole reference data allows the spatial extrapolation of many soil engineering properties, including ground ice characteristics, and has proven to be the most reliable mapping technique for large engineering projects (Jorgenson et al., 1998; Kreig, 1977). When used in conjunction with borehole information, ground-based geophysical surveys that included DC (direct current) resistivity,[6] capacitive-coupled resistivity, and GPR have also proved useful for mapping surficial materials and permafrost (De Pascale et al., 2008; Moorman et al., 2003). AEM surveys also provide information on subsurface stratigraphy, but interpretation is dependent on adequate borehole reference data and surficial geology (Minsley et al., 2012; Van Dam, 2012). Progress has been made with airborne GPR for subsurface mapping (Catapano et al., 2012). While photo-interpretation and integrated-terrain-unit mapping remains a practical and effective approach for local-scale mapping, particularly for engineering projects, emerging approaches that integrate automated landform characterization of DEMs, satellite imagery, and airborne geophysics with spatial statistical techniques show good potential for mapping large areas (Ho et al., 2012; Pastick et al., 2013). Many workshop participants indicated that 1-5 m is desired for local scales, but 100-1,000 m is sufficient on the circumpolar scale.

Soil physical properties. In particular, mineral composition, bulk density, and texture are critical to active layer dynamics and permafrost characteristics because they affect moisture-holding capacities, thermal properties, ice segregation, heave, and thaw settlement characteristics (Farouki, 1981). Traditionally, aerial photography and optical satellite imagery have been used to manually delineate homogeneous soil-landscape units for which soil properties have been established through field sampling. Soil properties have been mapped indirectly with optical and microwave data using physically based and empirical methods, including mineralogy, texture, soil iron, soil moisture, soil organic carbon, soil salinity, and carbonate content (Anderson and

[5] Epigenetic permafrost is formed in soils that have already been deposited by wind, water, and/or gravity.

[6] Resistivity is a quantification of how strongly a specific material opposes the flow of electric current. A material that is considered to have low resistivity readily allows the movement of electric charge.

Croft, 2009; Barnes et al., 2003; Mulder et al., 2011). Most investigations have used optical imagery, such as SPOT HRV and AVIRIS hyperspectral data, for quantifying bare soil properties or inferring soil properties from vegetation spectral responses (Barnes et al., 2003). Digital mapping methods are being developed using remotely sensed imagery for spatial interpolation of sparsely sampled soil properties (Browning and Duniway, 2011; Morris et al., 2008). Recently, Pastick et al. (2013) have developed machine-learning regression tree models using Landsat imagery, AEM surveys, and more than 20 ancillary layers to map active layer thickness and permafrost distribution in central Alaska. Workshop participants said that 1-5 m is ideal for local scales, but 100-1,000 m is sufficient on the circumpolar scale.

Soil organic layer. Properties such as thickness, moisture content, density, and thermal conductivity are important to permafrost dynamics because of their strong controls on soil thermal properties, active layer dynamics, and permafrost stability (Farouki, 1981; Johnson et al., 2013). Peat thickness and stratigraphy have been quantified through ground-based electrical geophysics (Slater and Reeve, 2002) and GPR (Laamrani et al., 2013; Rosa et al., 2009). Organic thickness or carbon contents have been related to hyperspectral remote sensing and field spectroscopy (Gomez et al., 2008; Jarmer et al., 2010). Spatial modeling using numerous terrain attributes (e.g., gridded climate data, terrain attributes derived from a DEM, land cover derived from Landsat) has been used to predict organic-layer thickness across Alaska (Mishra and Riley, 2012). Use of AEM, induced polarization imaging from SAR, and spatial modeling techniques that incorporate a wider range of optical and microwave satellite imagery deserve more investigation. High-resolution LiDAR can be used for detecting change in organic layer thickness over time and could provide useful information, especially before and after fire, said some participants. Soil organic layer has also been successfully inferred from Landsat-based indices and regression of field measurements of peat depth in drained thermokarst basins, allowing extrapolation of basin age as well as peat depth to some degree (Jones et al., 2012).

Moss characteristics. Properties such as moss thickness, moisture content, and thermal conductivity are similar to organic layer properties in their importance to understanding permafrost distribution and dynamics. Spectral reflectance characteristics have been used to differentiate some moss species, or species groups, that can be used to infer differences in hydrology and carbon cycling (Bubier et al., 1997). In patchy moss environments in Antarctica, a UAV equipped with a high-resolution camera and 6-band multispectral sensor enabled integration of high-resolution DEMs with a spectral classification to map moss mats (Lucieer et al., 2012). The passive microwave radiometer AMSR E/2 may have some potential, but its large footprint sizes (down to 6 × 4 km) may restrict its application to highly variable moss mats. Remote sensing of moss thickness in boreal and arctic ecosystems, however, has not yet been adequately demonstrated. A number of workshop participants said that a vertical resolution of 5-10 cm would be ideal pre- and post-disturbance. Some participants noted that HyspIRI's VSWIR (visible shortwave infrared) instrument will be useful for measuring moss characteristics, because it will provide a means for identification and classification of Arctic and subarctic vegetation.

Permafrost carbon. Organic matter preserved in frozen soils affects soil thermal properties and thaw settlement characteristics, and it is an important attribute of permafrost soils in which the large reservoir of frozen carbon can be released to the atmosphere after thawing, serving as a positive feedback loop to global climates (Jorgenson et al., 2013; McGuire et al., 2009; Schuur et al., 2008; Tarnocai et al., 2009). Traditionally, soil carbon contents have been associated with soil classification systems and related to terrain characteristics for mapping carbon stocks in the upper 1 m or more at coarse scales (Jones et al., 2010; Ping et al., 2011). Spatial modeling has been used to empirically model soil carbon distribution using geographic information systems (GIS) and thematic map inputs (Mishra and Riley, 2012; Zhou et al., 2008). Although some progress has been made in quantifying surface organics through field geophysical surveys (Laamrani et al., 2013), some workshop participants said that prospects for developing direct remote sensing techniques for quantifying soil carbon deep in ice-rich permafrost are poor. Many workshop participants indicated that 1-5 m is desired for local scales, but 100-1,000 m is sufficient on the circumpolar scale.

Soil salinity. The effect of soil salinity on the unfrozen water content of permafrost is important for understanding the thermal and structural properties of permafrost (Farouki, 1981). Most progress in remote sensing and geophysical techniques, however, has focused on salinization in arid regions (Farifteh et al., 2006). Optical satellite imagery has long been used for mapping and monitoring salt-affected soils and has been most effective in severely saline areas (Farifteh et al., 2006). Spectral indices have been developed for mapping salinity levels using EO-1 Hyperion hyperspectral imagery (Weng et al., 2010). AEM has been successful for mapping areas where highly conductive saline water has infiltrated geologic materials that have naturally low conductivities (Corwin, 2008). Areas with sufficient salinity to substantially affect permafrost distribution are mostly restricted to coastal areas; thus, development of new technologies and methods may not be necessary in the near term.

Floating mats. Floating mats often occur around the margins of thermokarst lakes and in thermokarst fens, which can lead to inaccuracies in the quantification of permafrost degradation associated with thermokarst lake development (Jones et al., 2011; Jorgenson et al., 2012). Floating mats or shore fens are easily mapped through photo-interpretation (Jorgenson et al., 2012). Remote sensing of spectral characteristics has been used to map floating mats and submergent vegetation (Cho et al., 2008). GPR has been used to determine the thickness and lateral expansion of floating mats and peat (Loisel et al., 2013; Parsekian et al., 2011). Given the restricted distribution of floating mats, many participants considered advancement of mapping technologies for these features not to be essential, although a horizontal resolution of 1-5 m is desirable.

Hydrology

Surface water bodies. Many workshop participants noted that thermokarst lakes and ponds are important indicators of permafrost degradation. They cover very large regions and have a significant impact on hydrology, geomorphology, and biogeochemical cycling in permafrost lowlands (Grosse et al., 2012). Several studies have used Landsat imagery (~15-80 m spatial resolution) to investigate changes in thermokarst lake/pond areas and relate these to either permafrost degradation or changes in precipitation/evapotranspiration regimes (e.g., Arp et al., 2011; Hinkel et al., 2007; Kravtsova and Bystrova, 2009; Labrecque et al., 2009). Although Landsat data are demonstrated to be useful for general change detection of lakes, they are not as effective for studying shore-erosion rates, which are usually too small for Landsat pixel size. SAR satellite imagery have been successfully used to map the extent of open water and, if available over an extended time period, can be useful for monitoring the extent and changes of surface water (Whitcomb et al., 2009). Aerial photographs and high-resolution satellite time series (~2.5 m spatial resolution or better; e.g., SPOT panchromatic, Ikonos-2, GeoEye-1, QuickBird, WorldView-1 and -2) are more suitable and commonly used to assess local and regional changes in thermokarst lakes/ponds in greater detail, including lake disappearance from drainage or drying and lake expansion by thermal erosion (Jones et al., 2011; Westermann et al., in press). Satellite measurements of water depth from green LiDAR[7] (e.g., Gao, 2009; Paine et al., 2013) and water level from near-IR LiDAR and radar altimeters (e.g., Crétaux et al., 2005; Paine et al., 2013) complement optical (Figure 2.6) and SAR observations of surface extent of lakes/ponds, but these have not been well explored to date. Workshop participants said that the future SWOT mission, to be launched in 2020, will provide invaluable data in this respect (e.g., Lee et al., 2010). The mission[8] will produce a water mask able to resolve lakes of 250 m^2 in size and will be able to retrieve water-level elevations with an accuracy of 10 cm.

In addition to water depth and water level, satellite imagery provides valuable information on shallow lakes (i.e., ~3 m or less) that freeze partially (i.e., sections) or entirely to their bed in winter. SAR data can be used to determine the areal extent and seasonal (i.e., wintertime) evolution of bedfast ice. Heat transferred

[7] LIDARs designed for mapping underwater use a blue-green laser that can penetrate water and provide returns of underwater objects or the bottom.

[8] See http://swot.jpl.nasa.gov/science/.

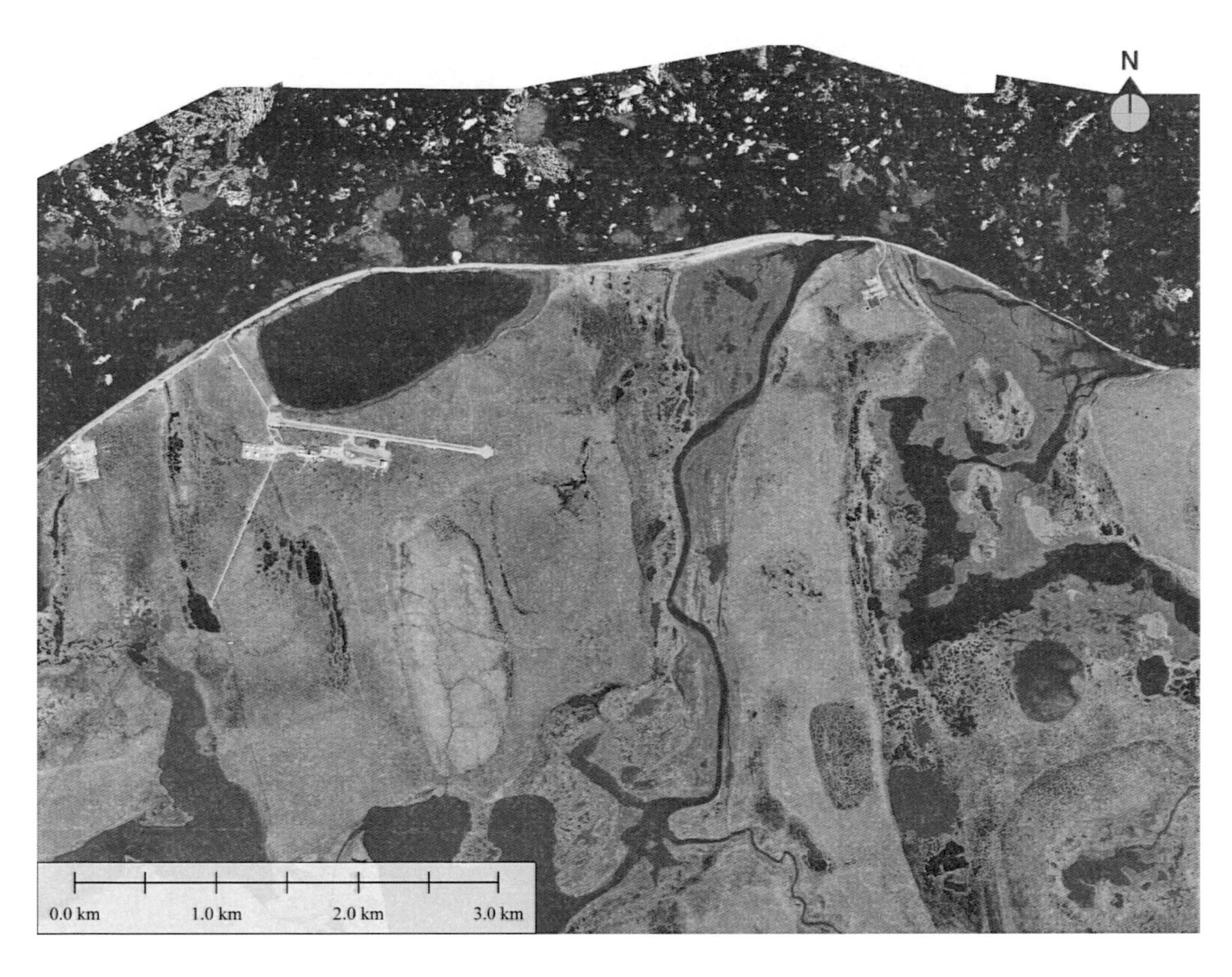

FIGURE 2.6 Historical optical data sets provide means for monitoring changing lakes near Lonely Air Force Station, Alaska. The source of the imagery is the Alaska High Altitude Aerial Photography (AHAP) program. The data were collected in July of 1979. SOURCE: Scott Arko, University of Alaska, Fairbanks.

from floating-ice lakes creates taliks. A shift from a bedfast-ice (i.e., frozen to bed) to a floating-ice regime can initiate talik development under the lake and potentially release large stocks of carbon previously frozen in permafrost in the form of methane (Arp et al., 2012). With Arctic climate warming, it is anticipated that a smaller number of lakes will freeze to their bed in winter because of a decrease in ice thickness (Surdu et al., 2013). Many workshop participants said that SAR data acquired at spatial resolutions of ~3-100 m and at weekly to monthly time scales are suitable for monitoring the evolution of floating ice and bedfast ice for lakes of various sizes.

Soil moisture. A significant terrestrial factor for controlling the surface energy balance after the presence or absence of snow cover is soil moisture. It is a major factor in permafrost aggradation and degradation because of the thermal properties of water and ice (Farouki, 1981). Soil moisture information is important for estimating the thermal properties of the ground needed for permafrost modeling. Information provided by SAR sensors is ideal for the spatial estimation or modeling of soil moisture at a range of different spatial scales. In contrast to passive microwave radiometers and scatterometers that provide surface soil moisture information at low spatial resolution (i.e., tens of km), some participants noted that SAR instruments (i.e., C, L, and P band; ~3-100 m resolution) are useful for medium- and large-scale analysis of soil moisture levels. However, there has been limited research to date on the estimation of relative or absolute surface soil moisture and its spatial distribution from SAR in permafrost regions. At subcatchment and finer scales in particular, the spatial patterns (e.g., heterogeneity, or relative spatial variations) of soil moisture become as or more important than the absolute value of the soil moisture at every point.

Many participants indicated that spaceborne or airborne microwave instruments that provide weekly acquisitions at spatial resolutions of 100 m or better and have a vertical sensitivity to the top 10 cm of the soil moisture profile (i.e., vertical resolution) would meet most user requirements. However, determination of soil moisture to greater depths (i.e., ~1 m into the ground; the top of the permafrost table) would be of even greater value to the permafrost community. SAR instruments operating at L-band and higher frequencies are best suited for the retrieval of surface soil moisture down to about 5 cm (Ulaby et al., 2013); those at P band or lower have the capability to penetrate deeper and allow retrievals of the soil moisture profile to depths of tens of centimeters or even more than 1 m, depending on soil texture composition and moisture content (Ulaby et al., 2013). No available P-band SAR instruments are operating from space, and the higher frequency spaceborne SARs have only sporadically been used for soil moisture retrievals. The upcoming NASA Soil Moisture Active/Passive (SMAP) mission (Entekhabi et al., 2010) is the first combined radar-radiometer spaceborne instrument set with the primary goal of mapping global distributions of soil moisture, including those in permafrost ecosystems (Figure 2.7). The SMAP soil moisture products will be at 3-, 9-, and 36-km scales, necessitating further disaggregation to arrive at higher resolution products. Airborne P-band radars, such as the AirMOSS system, are demonstrating the strong utility of such sensors for the retrieval of soil moisture profiles over a wide range of biomes at high resolutions (i.e., 90 m or better) and are expected to be highly suitable for doing the same over permafrost regions (Figure 2.8).

Some participants also indicated that the TIR (thermal infrared) instrument, which is planned as part of NASA's HyspIRI mission, would be useful for mapping surface soil moisture at medium resolution.

FIGURE 2.7 Image of the SMAP spacecraft structure. SOURCE: Presented by Dara Entekhabi.

Subsurface water (storage and flow). Several recent studies have shown a relationship between mass change as detected by the Gravity Recovery and Climate Experiment (GRACE) satellite mission and changes in terrestrial water storage (TWS) in the catchments of large Arctic rivers (Muskett and Romanovsky, 2009, 2011). In the Lena River basin, for example, observed TWS increase has been attributed to an increase in subsurface water storage between 2002 and 2010 (Velicogna et al., 2012). Missions such as GRACE may therefore allow a direct quantification of large-scale changes of the water balance triggered by permafrost thaw (Westermann et al., in press).

Water temperature. Lake water temperature influences talik development, thermal erosion, and ultimately surface permafrost degradation (Arp et al., 2010). Water temperature can be measured in situ with a radiometer that records "skin" temperature or, as done more frequently, using either a temperature sensor deployed just below the water surface or a set of temperature sensors installed on a chain for temperature measurements along a depth profile. Thermal sensors aboard airborne and satellite platforms provide measurements of surface

FIGURE 2.8 The AirMOSS sensor is located within the pod under the NASA Gulfstream III. SOURCE: Mahta Moghaddam, Principal Investigator, on behalf of AirMOSS EV-1 mission team.

"skin" water temperature or ice/snow temperature when ice cover is present on a lake. Current satellite sensors such as AVHRR and MODIS offer wide spatial coverage and high temporal resolution (twice daily or better) but sacrifice spatial resolution (~1 km), which allows the surface water temperature of only the largest lakes to be monitored. Landsat and the ASTER thermal sensors provide a better spatial resolution (i.e., ~100 m) than MODIS and AVHRR, but they have much reduced temporal sampling (e.g., Landsat only about once a week at best at high latitudes under clear sky conditions). In order to monitor the temperature of water bodies in permafrost regions (e.g., thermokarst lakes), satellite acquisitions with spatial resolutions of the order of ~30-100 m and daily revisits would be optimum. Participants discussed the Japanese satellite GCOM-C, which will be launched in the near future (2014) and will provide surface water temperature observations at 500 m spatial resolution that will allow for monitoring of lakes smaller than currently possible with AVHRR and MODIS.

Vegetation and Land Cover

Vegetation properties (e.g., cover, composition, structure, biomass) and processes (e.g., productivity) are important for permafrost research and mapping applications as indicated by several participants, because they provide essential indicators of permafrost presence (or absence), surface hydrology, and other key variables (e.g., surface temperature, snow cover and depth, active layer dynamics). The properties discussed below are useful for permafrost mapping applications.

Land cover. Vegetation characteristics and land cover are critical to assessing active layer and permafrost dynamics because they affect the surface energy budget, soil organic matter and thermal properties, evapotranspiration and water balance, and snow cover (Jorgenson et al., 2010). Numerous vegetation properties and processes have been remotely sensed using optical, radar, and LiDAR approaches, and there is extensive literature on sensors and methods (Belshe et al., 2013; Jones and Vaughan, 2010). Many aircraft instruments are suitable for vegetation mapping, including LiDAR, radar instruments, and even multifrequency digital camera mapping systems. Similarly, several space-based satellite missions are relevant to vegetation mapping, ranging from multiresolution optical sensors on a wide range of platforms (e.g., QuickBird, SPOT, Landsat, MODIS, AVHRR) to various radar missions (e.g., ALOS, ENVISAT, RADARSAT). Although participants noted that there is no single best system to map vegetation, a great deal has been accomplished in support of vegetation classification schemes relevant to permafrost landscapes using optical imagery alone, from simple classification schemes focused on structure (Selkowitz and Stehman, 2011) to ecologically diverse schemes (Jorgenson et al., 2009). Landsat and higher resolution imagery are particularly useful in this regard. The synergistic use of optical, radar, and optical remote sensing can provide a range of advantages in measuring some of the other desired vegetation variables (Selkowitz et al., 2012).

Vegetation structure and composition. Vegetation structure and composition are measured by height, species or lifeform density, leaf area index, and species biomass or cover, and they are important both to the surface energy balance and as indicators of permafrost degradation (Trucco et al., 2012). Optical imagery has been frequently used to produce vegetation indices related to productivity, particularly the Normalized Difference Vegetation Index (NDVI), to assess areas of change in Arctic and boreal regions (Beck et al., 2011b; Verbyla, 2008). Higher resolution time series of systematic vegetation indices would be useful for linking productivity changes to variability in permafrost properties decoupled from more regional-scale climate change, including those induced by fire disturbance. Coupling these higher resolution image series with LiDAR would be particularly useful for observing changes in vegetation structure and composition (e.g., Goetz et al., 2012; Wulder et al., 2007). Many workshop participants indicated that a vertical resolution of 5 cm and a horizontal resolution of 10-30 m are desirable.

Biomass. Biomass affects energy balance and soil carbon dynamics that are important to permafrost, and techniques for remotely sensing biomass are well developed (Frolking et al., 2009). Biomass has been estimated in permafrost environments with Landsat imagery (Ji et al., 2012), MODIS (Blackard et al., 2008), AVHRR (Raynolds et al., 2012), SAR (Thurner et al., 2013), InSAR (Solberg et al., 2013), and airborne and spaceborne LiDAR (Neigh et al., 2013). Estimation using multisensor approaches synergistically helps to overcome the limitations of each of these systems used independently (whether due to inadequate LiDAR sampling, surface moisture effects in radar backscatter, or pervasive cloudiness in optical imagery). Biomass indices in particular have proven to be significant factors in spatial modeling of permafrost properties (Pastick et al., 2013).

Disturbance. Fire, geomorphic processes, and human activities affect active layer and permafrost dynamics by altering the surface microclimate related to vegetation, the organic-layer properties, and surface hydrology (Brown and Grave, 1979; Yoshikawa and Hinzman, 2003). Fire is of particular importance because it is widespread in boreal ecosystems and becoming more frequent, even in tundra ecosystems (Kasischke et al., 2010). Fire severity, which affects recovery patterns and depth of organic-layer combustion, is critical to assessing permafrost response (Harden et al., 2006; Nossov et al., 2013). A number of remote sensing indices using Landsat imagery have been used to quantify fire severity (Hoy et al., 2008). Recently, fire boundaries were used in spatial modeling to map active layer depths and permafrost distribution in central Alaska (Pastick et al., 2013). Landsat imagery, which now includes the operational Landsat 8, provides adequate resolution, coverage, and frequency to assess the relationship of fire disturbance and permafrost stability.

The primary planned future sensors that will be useful for vegetation mapping include the U.S.

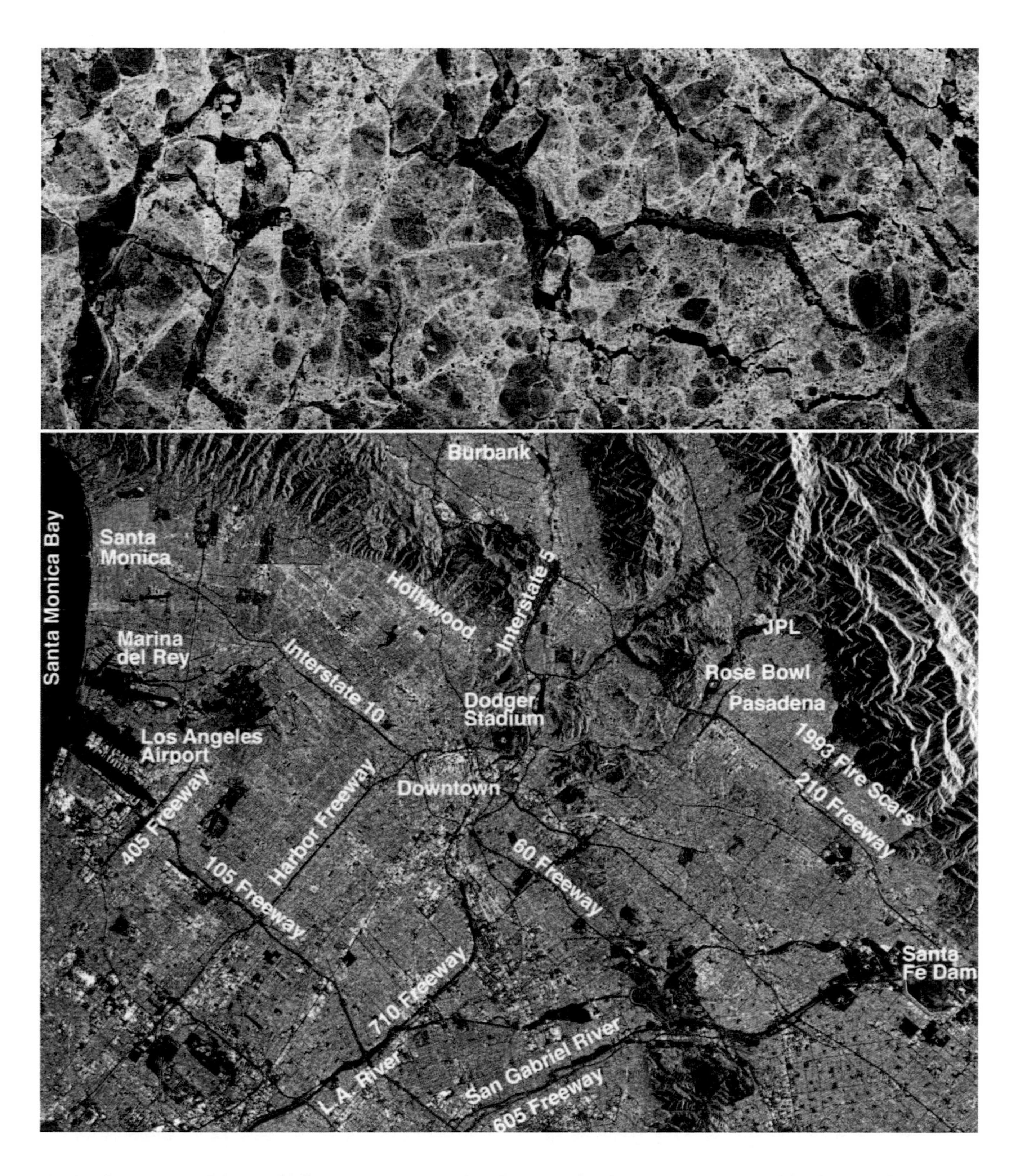

FIGURE 2.9 Images of the Weddell Sea, Antarctica (top) and Los Angeles (bottom), were acquired by the Spaceborne Imaging Radar-C/X-Band Synthetic Aperture Radar (SIR-C/X-SAR) onboard the space shuttle Endeavour. These images highlight the potential benefits of future L-band and S-band sensors for advancing permafrost research. For example, a combination of simultaneous S-band and L-band data would be extremely powerful for discriminating differential scales. Greater available bandwidth at S-band than at L-band could enable focus on some areas at finer resolution. Colors are assigned to different radar frequencies and polarizations as follows: red is L-band horizontally transmitted, horizontally received; green is L-band horizontally transmitted, vertically received; and blue is C-band horizontally transmitted, vertically received. SOURCE: NASA.

L-band InSAR mission (Figure 2.9), the series of European Sentinel missions, and the Japanese ALOS-2 PALSAR and GCOM-C missions, among others. The utility of the photo counting ATLAS sensor planned for ICESat-2 for vegetation mapping is currently uncertain because of the very few photons returned within any given fine-scale grid cell. A future LiDAR mission to provide coincident vegetation cover, structure, and other high-resolution surface variables, such as subcanopy topography, is considered desirable by many workshop participants.

Greenhouse Gases

In the climate research community, much of the concern about changes in permafrost actually relate to its role in the global carbon cycle, including uncertainty about whether permafrost-dominated regions will become net sinks or sources of greenhouse gases as permafrost thaws. A number of workshop participants said that the importance of these issues means that satellite measurements of carbon dioxide (CO_2), methane, and water vapor should be a top priority. Measurements of CO_2 and/or methane have been made with the JAXA GOSAT (still flying as of November 2013) and the ESA SCIAMACHY (mission now over) satellites/sensors, as well as via aircraft (NASA's CARVE and other programs; Zulueta et al., 2011) and ground-based towers. A number of future carbon sensors are proposed by NASA and other agencies that, if built, will have the appropriate resolution to detect changes in carbon fluxes from thawing permafrost (100 m to 1 km).

Methane fluxes. Methane fluxes in surface water bodies can also be inferred from bubbles trapped in lake ice covers, which can be detected by C- and L-band SAR from satellite or aircraft (Engram et al., 2012; Walter et al., 2008). U.S. researchers, however, have limited access to RADARSAT-2 data, the only current C-band SAR in orbit. Historical data from RADARSAT-1, ERS-1 and -2, Envisat ASAR, and ALOS-1 are still valuable for research. As discussed at the workshop, ALOS-2 may provide useful L-band data for methane detection in lake ice after it is successfully launched.

Water vapor. Another important greenhouse gas related to permafrost is water vapor. Clouds are ubiquitous in Arctic imagery and are monitored in detail by the meteorological satellites in the GOES and POES series. Passive microwave sensing is used to detect water in its vapor form before it has condensed into clouds. The origin of water vapor can, like carbon, be traced to some extent using isotopes. Deuterium is measured by the Tropospheric Emissions Spectrometer onboard the Aura satellite and can be used to trace the large-scale movement of water vapor parcels, globally. Water vapor isotopes have also been successfully measured in situ from aircraft and towers. Current product resolutions probably suffice for remote sensing of large-scale water vapor; however, estimates of water vapor fluxes between permafrost-dominated landscapes and the atmosphere in the form of evapotranspiration are limited to a few in situ measurements.

Carbon storage. Indications of changes in carbon storage can include vegetation change leading to "greening" and "browning" trends, as seen in measures such as NDVI. The pixel resolution for NDVI depends on the sensor that is being used: derived from Landsat it is tens of meters and from MODIS 500 m. Changes in vegetation type that may affect carbon budgets (such as shrub expansion) are likely occurring on scales of 1-10 m per decade, suggesting that Landsat ETM may be a better source for NDVI for this application. On the other hand, MODIS's daily repeat cycle makes it better for looking at seasonal shifts in larger-scale NDVI. Workshop participants indicated that an ideal sensor would have a daily repeat cycle with 1-10 m horizontal resolution.

^{13}C and ^{14}C. The question of carbon source region and age are best answered by ^{13}C and ^{14}C isotopes, which are currently not measured by satellite but more typically through lab analysis of in situ samples. This would be a useful measurement from remote sensing, but according to many workshop participants would be less so than carbon flux measurements.

DATA FUSION

The workshop discussion focused on analysis of existing sensors or those on the horizon rather than the development of new analytical methods utilizing existing data. However, assimilation of multiple

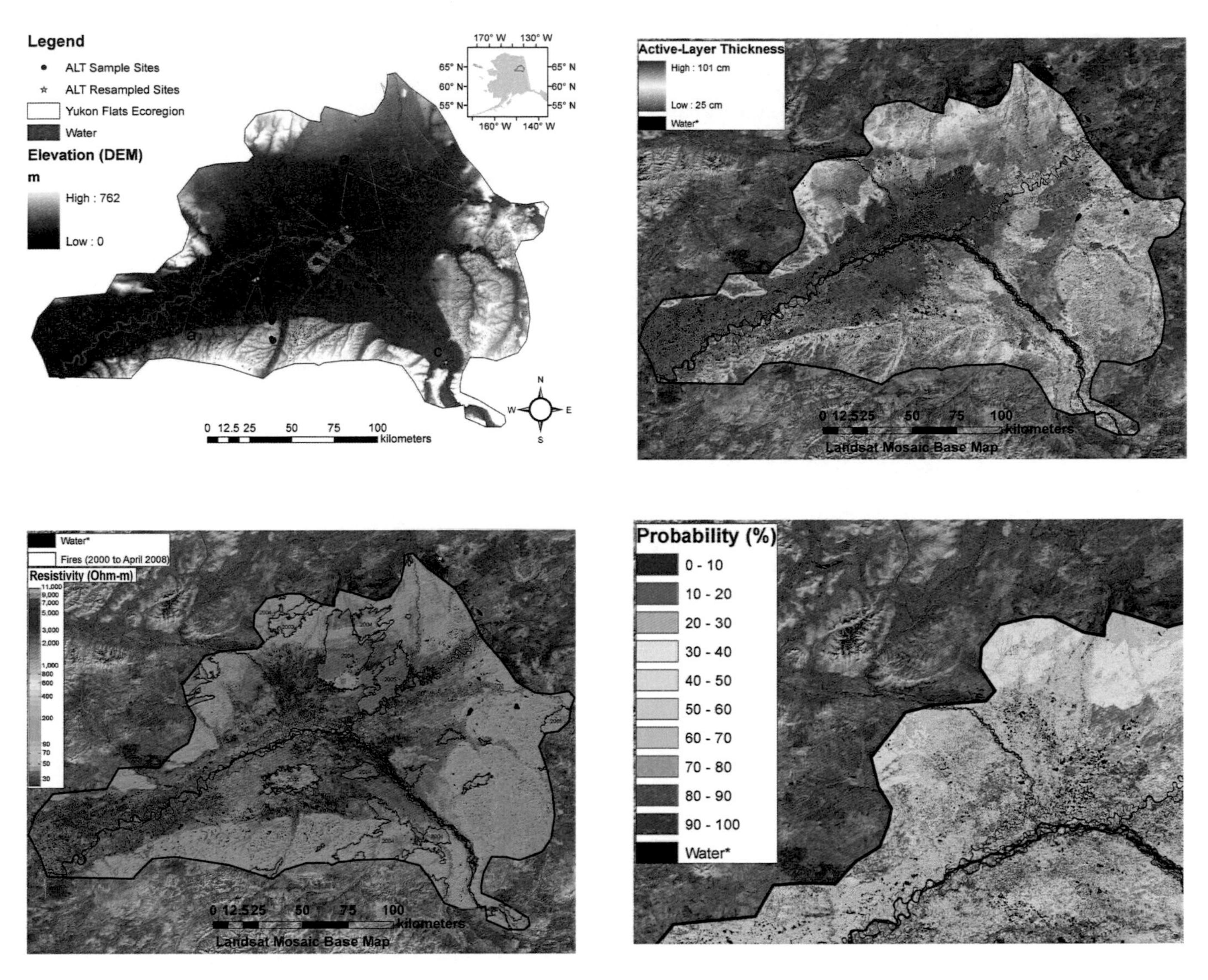

FIGURE 2.10 Shallow AEM data (top left) combined with satellite and auxiliary data sets to extrapolate near-surface resistivity (top right); predict ALT (bottom left); and estimate probability of shallow permafrost (bottom right). SOURCE: Pastick et al., 2013.

data sources for creation of new products could help to address some of the information gaps. Several participants noted that the fusion of remotely sensed data from multiple sensors, field measurements, and improved temporal and spatial modeling is essential to making progress in mapping permafrost and detecting change (Figure 2.10). Because permafrost responds to a wide range of ecological factors and is covered by surface vegetation and by soil within the active layer, it is increasingly apparent that no single sensor is capable of reliably mapping permafrost properties.

Although AEM shows substantial capabilities and ongoing innovations are promising, data acquisition and processing are intensive and costly, thereby limiting their use to local-scale mapping projects (Minsley et al., 2012). AEM data can be spatially extrapolated through multiparameter data fusion approaches incorporating Landsat imagery, DEM and terrain modeling, and surficial geology (Pastick et al., 2013). Approaches using large data stacks and multiparameter fusion, said some participants, should be extended to include remotely sensed field-calibrated SAR-derived surface soil moisture and active layer freeze thaw, SAR and optically derived snow properties, optically derived vegetation indices and vegetation classification, LiDAR-derived vegetation structure and topographic-moisture indices, spatial extrapolation of soil surface-organic-matter field data, and optically derived fire indices. The data can be fused through both empirical multivariate analyses (Mishra and Riley, 2012; Panda et al., 2012; Pastick et al., 2013) and processes modeling ground thermal regimes (Jafarov et al., 2012; Marchenko et al., 2008). Many workshop participants noted that incorporation of field and remotely sensed data to parameterize the process models will improve the temporal response and spatial resolution of input variables.

However, numerous participants indicated that a critical gap remains in landscape- to local-scale mapping of surficial geology, because substantial areas are mapped at only small regional scales, and most of the mapping is outdated with poor spatial accuracy that is incompatible with remotely sensed data. This mapping of subsurface materials and stratigraphic relations is critical for modeling subsurface thermal characteristics (Jafarov et al., 2012), mapping the distribution of ground ice (Jorgenson et al., 2008; Kanevskiy et al., 2011; Kreig, 1977), and understanding the patterns and processes of thermokarst (French and Shur, 2010; Jorgenson et al., 2013). Because remote sensing and automated mapping of surficial geology currently is not possible, it will require a substantial effort to compile field data from surficial geology investigations, compile and digitize existing mapping with updated concepts, and create new maps for extensive unmapped regions.

TABLES 2.1 AND 2.2 APPEAR ON THE FOLLOWING PAGES.

TABLE 2.1 Direct and Indirect Remote Sensing of Permafrost Properties and Processes

What variables need to be measured to study permafrost?		What remote sensing techniques can make these measurements?					
Permafrost properties and processes	Ecological indicators of this property	Current systems that can measure this property	Current systems that can measure ecological indicators	Prototype sensors and future possibilities	What spatial and temporal resolutions are available (including coverage)	What spatial and temporal resolutions are needed (including coverage)	Relevant references
			PERMAFROST				
Active layer thickness	Seasonal heave/ subsidence Physical changes within the active layer (including seasonal thawing and seasonal freezing)	AirMOSS (P-band SAR) UAVSAR (L-band InSAR) Airborne-EM OIB low-frequency radars	AirMOSS (P-band SAR) UAVSAR (L-band InSAR)	Multiband InSAR with LiDAR (as a correlate) Surface and airborne GPRs AirMOSS (P-band SAR) in InSAR mode; simultaneous AirMOSS+UAVSAR (P- and L-band SAR) SPOT 1, HRV, MLA, AVHRR	LVIS airborne LiDAR: 3 m horizontal resolution UAVSAR: 10 m horizontal resolution and cm-level subsidence sensitivity AirMOSS: 30 m horizontal resolution and 5-10 cm vertical sensitivity AVHRR: 1 km	Vertical resolution: 5 cm Horizontal resolution 30 m Biweekly (except in winter) Current is adequate, especially if used in combination	Hubbard et al., 2013; McGuire et al., 2002; Minsley et al., 2012; Pastick et al., 2013; Peddle and Franklin, 1993
Ground ice (volume and morphology)	Thermokarst landforms Surface topography/ subsidence Dynamics of thermokarst lake margins Lakes bathymetry Surficial geology/ landforms	GPR	Quickbird, IKONOS, CORONA Airborne and space-based LiDAR High-resolution optical and SAR Declassified imagery	Direct detection of ground ice, remote sensing GPR, OIB Radars Some sort of fusion sensor system; e.g., resistivity and hyperspecral AEM EM modeling OIB Radars Airborne InSAR	1-10 m (commercial satellite) up to daily demand <1 m airborne SAR and GPR on demand	1-10 m is sufficient Vertical: <0.5 m 1 m with InSAR is achievable and would be useful Annually Need to clearly define the application (e.g., engineering, ecological)	Grosse et al., 2005; Jones et al., 2011; Kreig, 1977; Sannel and Brown, 2010; Yoshikawa et al., 2006
Subsurface soil temperature	Land cover type (weakly linked)	None	None	Something that provides profile temperature, subsurface sampling down to depth of 0 annual amplitude Low-frequency radiometers	N/A	N/A	
Permafrost presence/ absence	Land surface temperature Vegetation Landform	EM AEM	Thermal and passive microwave Various optical	Space and airborne radars	Annually	Local 1-10 m Circumpolar 100 m Annually	Minsley et al., 2012; Panda et al., 2010

Depth to top of permafrost	Poorly understood	AEM EM modeling Multifrequency airborne low-frequency radars (OIB)	None	Low-frequency ultra-wideband radars	None	Local 1-10 m Circumpolar 100 m Annually	Annan and Davis, 1976; Arcone et al., 1998; Brosten et al., 2006; Rodriguez-Morales et al., 2013
Near-surface seasonal freeze/thaw state	Presence/absence of liquid water Seasonal heave/ subsidence Plant phenology Presence of snow	InSAR, repeat LiDAR, RADARSAT Tomographic SAR AEM GPR Times series of radar/ radiometer SMAP[a]	NASA aircraft SARs UAVSAR, AirMOSS Spatial extrapolation based on vegetation coverage Higher frequency passive and active	High-frequency repeat satellite LiDAR. InSAR ALOS2, NASA L-band SAR (former DESDynI)[b] space based & airborne GPR & EM multifrequency active passive microwave sensors	UAVSAR: 10 m AirMOSS: 30 m SMAP: 3 km to 36 km (3-day repeat) ALOS2: 15-30 m (~2-3-week repeat)	10 min in every 10 days Vertical: 5-10 cm Subsidence < 1cm Local: 1 m Circumpolar: 100 m	Park et al., 2010
Permafrost thickness and 3D geometry	Lakes Land cover type (weakly linked)	AEM GPR(VHF and UHF) Tomographic SAR in VHF and P-band	Any optical sensors	Satellite-based EM GPR P-band SAR OIB VHF and UHF radars	Horizontal: 50-300 m Vertical: 2-20 m	Improved <2 m near-surface resolution Vertical: 10 m Horizontal: 10 m	Minsley et al., 2012

[a] Proposed launch date for late 2014 (NRC, 2012).
[b] No proposed launch date, but the NRC recommended a 2010-2013 launch time frame (NRC, 2012).

TABLE 2.2 Direct and Indirect Remote Sensing of Permafrost-Related Ecological Variables

What variables need to be measured to study permafrost?		What remote sensing techniques can make these measurements?					
Permafrost-related ecological variables	Ecological indicators of this variable	Current systems that can measure this variable	Current systems that can measure ecological indicators	Prototype sensors and future possibilities	What spatial and temporal resolutions are available (including coverage)	What spatial and temporal resolutions are needed (including coverage)	Relevant references
CLIMATE							
Snow characteristics (extent, water equivalent, depth density, conductivity)	Vegetation structure Topography Climatic parameters Land cover type (weakly linked)	High-frequency (Ku) SAR backscatter Passive microwave (e.g., AMSR, SSM/I) GRACE, optical (extent) or optical/ microwave combined InSAR, LiDAR Snowradar GPR	LiDAR Stereoimages D-InSAR, extrapolation GPS (signal delay), optical/microwave combined Spatial extrapolation SNOTEL	Ultra wideband radar High-frequency/ multifrequency SAR (e.g., CoReH2O) GRACE-FO[a] New optical sensors (extent)	Airborne ~1 m Passive microwave ~25 km Daily	Horizontal: 1-50 m Daily	Derksen et al., 2003; Hall and Riggs, 2007; Hancock et al., 2013; Kelly et al., 2003
Land surface temperature	Snow Plant phenology (longer time scales)	AMSR and SSM/I (snow-free conditions) MODIS SNPP Aircraft AVHRR Landsat ASTER airborne HyspIRI	AMSR and SSM/I (snow-free conditions) MODIS SNPP Aircraft AVHRR Landsat ASTER airborne HyspIRI	Landsat AVHRR and MODIS continuation Sentinel-3 GCOM-C HyspIRI satellite[b]	Horizontal resolution: 90 m ASTER 100 m Landsat-8 1 km MODIS/ AVHRR 25 km AMSR and SSM/I Twice monthly to daily	Horizontal: 30-100 m Daily	Hachem et al., 2009; Jones et al., 2010; Kimball et al., 2009
Meteorology (including air temperature, precipitation, winds)	Land cover Topography (longer time scales)	GOES POES AIRS on Aqua CrIS on SNPP AMSR	N/A	Continuation of meteorological satellite constellations Improve measurement of solid precipitation	Horizontal: 500 m to 100 km Hourly to daily	Horizontal: 50 m to 1 km Hourly to daily	Liu et al., 2006; Revercomb et al., 2013

				Global Precipitation Measurement (GPM) mission (2014)			
TOPOGRAPHY							
Surface topography (static, macro-, and micro-)	Land cover type Drainage patterns Water bodies Landforms Patterned ground	LiDAR Tandem-X (World DEM) LiDAR/SAR synergy	Various/multisensor	US L-band SAR DLR-JAXA Tandem-L (proposed) ICESat[c]-ATLAS	Vertical: <15 cm Horizontal: 1 m	Horizontal: <1 m Vertical: 5-10 cm Snow free: yearly	Bowen and Waltermire, 2002; Jones et al., 2013
Longer-term surface subsidence	Land cover type Drainage patterns Water bodies Landforms Patterned ground	LiDAR Tandem-X (World DEM) LiDAR/SAR synergy	Various/multisensor	US L-band SAR DLR-JAXA Tandem-L (proposed) ICESat-ATLAS	Vertical: <15 cm Horizontal: 1 m	Horizontal: 1 m Vertical: 1-10 cm Snow free: yearly	Chen et al., 2013; Liu et al., 2012
Thermokarst distribution	Long-term surface subsidence Thermokarst landforms Thermo-erosional features and slope failures Land cover type, water bodies Drainage patterns Patterned ground	Airborne LiDAR High-resolution optical Time series of SAR and InSAR images AEM	Various/multisensor	US L-band SAR DLR-JAXA Tandem-L (proposed) ICESat-ATLAS	Vertical: <15 cm Horizontal: 1 m	Horizontal: 1 m Vertical: 1-10 cm Snow free: yearly	Jones et al., 2012
Seasonal heave/ subsidence	Patterned ground Infrastructure damage	LiDAR InSAR	Various/multisensor	US L-band SAR ESA Tandem-L ICESat-ATLAS	Vertical: <15 cm (airborne) Horizontal: 1 m	Vertical: 5-10 cm Subsidence < 1cm Local scale: 1 m Circumpolar: 100 m per decade	Short et al., 2011

Continued

TABLE 2.2 Continued

What variables need to be measured to study permafrost?		What remote sensing techniques can make these measurements?					
Permafrost-related ecological variables	Ecological indicators of this variable	Current systems that can measure this variable	Current systems that can measure ecological indicators	Prototype sensors and future possibilities	What spatial and temporal resolutions are available (including coverage)	What spatial and temporal resolutions are needed (including coverage)	Relevant references
GEOLOGY AND SOIL							
Surficial geology-terrain units (including lithology, bedrock)	Topography Vegetation Drainage patterns Landforms	AEM GPR	Photo-interpretation of stereo airphotos and satellite imagery (e.g., Quickbird, Ikonos) Stereo-photogrammetry LiDAR	ALOS-2 Stereo capability Color CIR, Airborne GPR	<1 m for airphotos <1 m panchromatic and color satellite Horizontal: ~50-300 m Vertical: ~2-20 m	Local scale: 1-5 m Circumpolar: 100-1000 m	Catapano et al., 2012; Kreig and Reger, 1982; Minsley et al., 2012; Moorman et al., 2003; Mulder et al., 2011; Rawlinson, 1993; Siemon, 2006
Soil organic layer (thickness, moisture, conductivity)	Topography Vegetation Drainage patterns landforms	GPR EM resistivity Induced Polarization Imaging	Spatial extrapolation from soil cores (data fusion Landsat, ASTER, DEM etc.) LiDAR	Space borne, airborne sensor (like GPR)	<1 m for GPR 30-250 m resolution for optical satellites	Vertical: 5-10 cm pre- and post-disturbance	Gomez et al., 2008; Jarmer et al., 2010; Johnson et al., 2013; Laamrani et al., 2013; Mishra, 2013; Rosa et al., 2009; Slater and Reeve, 2002
Moss layer (thickness moisture content, thermal conductivity)	Vegetation type Topography Drainage patterns Landforms	N/A	AVIRIS CASI AMSR E/2 LiDAR	HYspIRI	1-5 m airborne	Vertical: 5-10 cm pre- and post-disturbance (LiDAR, InSAR)	Bubier et al., 1997; Lucieer et al., 2012
Permafrost carbon (deep frozen organic matter)	Surficial geology Landforms Vegetation Carbon isotopes	N/A	Spatial extrapolation from soil cores (data fusion Landsat, ASTER, DEM etc.)	Not discussed	100-1000 m	1-5 m local scale, 100-1000 m circumpolar	Jorgenson et al., 2013; Mishra, 2013; Ping et al., 2011; Tarnocai et al., 2009

Soil physical characteristics (composition, structure, density, texture)	Surficial geology Topography Vegetation Drainage patterns	AEM	Spatial extrapolation from soil cores (data fusion Landsat, ASTER, DEM etc.) Hyperspectral AVIRIS	Not discussed	1-5 m (airborne)	Local scale: 1-5 m, Circumpolar: 100-1000 m	Anderson and Croft, 2009; Barnes et al., 2003; Browning and Duniway, 2011; Hively et al., 2011; Morris et al., 2008; Mulder et al., 2011
Unfrozen/saline water mapping	Vegetation Topography	AEM	Optical spectral mapping of salt tolerant vegetation (Landsat, Ikonos)	Not discussed	Horizontal: 50-300 m Vertical: 2-20 m (for AEM)	Improved <2 m near-surface resolution	Corwin, 2008; Farifteh et al., 2006; Fitterman and Deszcz-Pan, 1998; Paine and Minty, 2005
Floating mats in water bodies (depth, width)	Water bodies Topography Vegetation type	Ground GPR	High-resolution optical satellite sensors Airphotos Hyperspectral	HYspIRI	1-5 m	Local scale: 1-5 m not applicable to circumpolar scale	Cho et al., 2008; Jorgenson et al., 2012; Loisel et al., 2013; Parsekian et al., 2011
HYDROLOGY							
Surface water bodies (including dynamics, redistribution)	Lake/ponds ice thickness Change in depth and extent of the water bodies on the land surface (lakes, other bodies) Lake drainage Changes in wetlands distribution	High to low resolution optical (e.g., GeoEye/ IKONO, QuickBird, WorldView, Landsat, SPOT, MODIS) SAR, green LiDAR (water depth), NIR, LiDAR , and radar altimetry (water level) SRTM (lake drainage)	High to low resolution optical (e.g., GeoEye/IKONOS, QuickBird, WorldView, Landsat, SPOT, MODIS) SAR, green LiDAR (water depth), LiDAR SRTM (lake drainage)	Continuation of high-resolution optical and SAR, SWOT,[d] Air-SWOT, HyspIRI	Horizontal: 0.5 m to 1 km Daily to monthly	Horizontal: Current spatial is sufficient Weekly	Arp et al., 2011; Hinkel et al., 2007; Kravtsova and Bystrova, 2009; Labrecque et al., 2009; Lee et al., 2010; Paine et al., 2013

Continued

TABLE 2.2 Continued

What variables need to be measured to study permafrost?		What remote sensing techniques can make these measurements?					
Permafrost-related ecological variables	Ecological indicators of this variable	Current systems that can measure this variable	Current systems that can measure ecological indicators	Prototype sensors and future possibilities	What spatial and temporal resolutions are available (including coverage)	What spatial and temporal resolutions are needed (including coverage)	Relevant references
Soil moisture	Surface soil moisture dynamics Soil moisture profile dynamics Depth of thawed layer Vegetation for spatial patterns DEM topographic indices for broad patterns	AirMOSS, AIRSAR (P- and L-band SAR, but obsolete) AMSR-E/2 C-band SAR Landsat SMOS PALS ASCAT, ASAR, SAR backscatter, microwave emission	UAVSAR, AirMOSS AirMOSS LST and NDVI	AirMOSS Combination of AirMOSS and UAVSAR (simultaneous P- and L-band SAR) Multifrequency radiometer NASA L-band SAR Mission (former DESDynI) GCOM-W (Global Change Observation Mission-Water) L-Band sensor SWOT HYspIRI	Potentially 3 m to 30-100 m resolution Airborne instrument, so any temporal resolution is possible Depth of sensing is expected to reach 50-100 cm with P-band SAR (AirMOSS) Pan-arctic 25 km MODIS NDVI daily ~500 m, Worldview2 NDVI, 1-10 m daily on demand, NASA aircraft SARs L-Band is coming on-line (limited penetration)	Local to regional scale: 1 m horizontal, daily Circumpolar: 100 m horizontal, weekly Vertical: 10 cm depth minimum; ideally to top of permafrost	Moghaddam et al., 2000; Liu et al., 2011; Tabatabaeenejad et al., in review; Tabatabaeenejad and Moghaddam, 2011
Subsurface water (storage and flow)	Drainage patterns Vegetation	GRACE GPS		GRACE-FO	Horizontal: : ~400 km Every 10 days		Muskett and Romanovsky, 2009, 2011; Velicogna et al., 2012
Water temperature	Water bloom Ice cover	MODIS SNPP Aircraft AVHRR Landsat ASTE airborne HYspIRI	AMSR and SSM/I (snow-free conditions) MODIS SNPP Aircraft AVHRR Landsat ASTE airborne HYspIRI	Landsat AVHRR and MODIS continuation Sentinel-3, GCOM-C HYspIRI satellite	Horizontal: 90 m (ASTER); 100 m (Landsat-8) 1 km (MODIS/ AVHRR) Twice monthly to daily	Horizontal: 30-100 m Daily	Arp et al., 2010; Kheyrollah Pour et al., 2012

VEGETATION AND LAND COVER							
Land cover (including spectral vegetation indices)	Vegetation type Water bodies Land use Dying forests NDVI (incl. dynamics)	Multispectral SAR Optical	Multispectral SAR Optical	Hyperspectral sensor High-resolution optical and SAR imagers	0.5 m (panchromatic) and above (WorldView2)	Current spatial is sufficient	Jones and Vaughan, 2010; Jorgenson et al., 2009; Selkowitz and Stehman, 2011
Vegetation structure and composition	Height Density LAI Species type	Aircraft LiDAR; SAR Optical (composition)	Hyperspectral	US L-band SAR DLR-JAXA Tandem-L (proposed) GCOM-C (composition)	Vertical: 15 cm (airborne) Horizontal: 1 m	Vertical: 5 cm Horizontal: <1 m; 10-30 m annual	Beck et al., 2011a; Goetz et al., 2010
Biomass (above ground)	Woody biomass Surface vegetation	Aircraft LiDAR SAR	Aircraft LiDAR SAR optical	US L-band SAR DLR-JAXA Tandem-L (proposed)	<1 m airborne <30 m satellite	Current spatial is sufficient seasonal	Blackard et al., 2008; Ji et al., 2012; Neigh et al., 2013; Raynolds et al., 2012
Disturbance (including history)	Metrics (e.g., severity) Fire, humans, flooding, insects	Optical Aircraft LiDAR SAR	Aircraft LiDAR SAR optical	Hyperspectral sensor High-resolution optical and SAR imagers	0.5 m (panchromatic) and above (WorldView2)	Current spatial is sufficient annual	Epting et al., 2005; Goetz et al., 2007; Hoy et al., 2008; Kasischke et al., 2010
GREENHOUSE GASES							
Methane (flux or concentration)	Vegetation Water bodies Lake ice	Airborne measurements (FTS) GOSAT SCIAMACHY (no longer in use) C-band SAR (methane trapped in ice)	C- and L-band SAR (bubbles)	OCO-2, OCO-3 FLEX (proposed) CarbonSat (proposed) Sentinel 5 precursor (proposed)	C-band SAR (30 m); L-band SAR (10 m); GOSAT, SCIAMACHY, OCO (>100 m)	100 m to 10 km	Engram et al., 2012; Walter et al., 2008
Water vapor flux	Vegetation Water bodies	Concentrations currently measured by satellite Water vapor isotopes	Concentrations currently measured by MODIS, GOES, POES, SNPP in infrared and isotopes by TES	Probably need product development rather than new sensors, need follow on to TES	10 km (model output)	10 m to 1 km	Groves and Francis, 2002a, b

Continued

TABLE 2.2 Continued

What variables need to be measured to study permafrost?		What remote sensing techniques can make these measurements?					
Permafrost-related ecological variables	Ecological indicators of this variable	Current systems that can measure this variable	Current systems that can measure ecological indicators	Prototype sensors and future possibilities	What spatial and temporal resolutions are available (including coverage)	What spatial and temporal resolutions are needed (including coverage)	Relevant references
CO_2 (flux or concentration)	Vegetation Water bodies NDVI	Airborne measurements (FTS) GOSAT	Chlorophyll, NPP inferred from multispectral, Landsat, MODIS	OCO-2, OCO-3 FLEX (proposed) CarbonSat (proposed) Sentinel 5 precursor (proposed)	1-100 m (imagery) 1-10 km (spectrometry)	1-100 m (imagery) 100 m to 10 km (spectrometry)	Zulueta et al., 2011
13C and 14C isotopes	Surficial geology (from landforms, vegetation)	None	RGB (MODIS, Landsat, SNPP)	Spectrometer	1-500 m (imagery only)	100 m to 10 km	

[a] No planned launch date; NRC recommended a 2016-2020 launch time frame (NRC, 2012).
[b] Currently in the study stage; the NRC recommended a 2013-2016 launch time frame (NRC, 2012).
[c] Proposed launch date for 2016.
[d] No proposed launch date; NRC recommended a 2013-2016 launch time frame (NRC, 2012).

3

Future Opportunities

It is clear from the workshop discussions that a multiscale, multisensor approach, using ground-based, airborne, and spaceborne instruments, would substantially advance the current state of knowledge of permafrost landscapes and, in the process, provide critically needed information on subsurface properties that determine the vulnerability of permafrost systems to warming. The data sets and products thus derived—and necessarily calibrated and validated with field measurements—would allow parameterization and development of more realistic permafrost models than would otherwise be possible. Taken together, these remote sensing–derived maps of properties and improved models would advance our understanding and prediction of the state of permafrost landscapes and associated feedbacks to the climate system.

Many participants said that additional high-quality information on permafrost properties is needed for models to realistically capture processes and faithfully predict likely future outcomes at any scale—from local to regional to global. To achieve this desired objective, they noted that a range of remote sensing data sets are needed, from measurements of properties of the land surface that provide indirect indicators of permafrost properties to more direct measurements that can be used to estimate those same properties. Specific desired temporal and spatial resolutions for each variable can be found in Tables 2.1 and 2.2. Many workshop participants noted that ice content is particularly important, because it strongly influences the vulnerability of permafrost to degradation (thaw) and thus the processes that follow that degradation (e.g., surface subsidence, thermokarst formation, and the release of old carbon frozen for decades to centuries and millennia).

Although there are a number of current remote sensing instruments on various platforms and planned missions for estimating surface variables relevant to indirect inference of permafrost properties, there are few such instruments or missions for direct estimation of permafrost properties. Several participants said that current remote sensing observations and derived data sets will be used in innovative ways to map aspects of permafrost landscapes or indirectly infer subsurface properties. The upcoming ICESat-2 mission will also be valuable for providing time series of surface topography that could be used to map surface deformation and thermokarst features.

However, as noted by numerous participants, there is a pressing need to advance remote sensing products in the near future, in particular for direct observations of permafrost properties. Polarimetric, InSAR, and LiDAR may be particularly valuable. Of the planned spaceborne missions, workshop participants considered the SMAP mission, planned for launch in late 2014, to be valuable for providing frequent (2-3 days) freeze/thaw and soil moisture products, albeit at relatively coarse (~3 km at best) spatial resolution. The U.S. L-band polarimetric InSAR mission would be particularly valuable for mapping higher resolution (100 m or better) seasonal freeze/thaw cycles, surface deformation, and subsidence. Even so, additional advancements are needed to more directly map permafrost features. P-band SAR, such as that planned for the BIOMASS mission, scheduled for launch in 2019, has the potential to advance remote sensing of active layer thickness and soil moisture content (and, based on studies with GPR, perhaps even ice content). However, BIOMASS P-band satellite will be restricted from transmitting in most of North America, Europe, and northern Eurasia because of spectrum usage conflicts (especially usage by military services). Such restriction is currently not imposed in Central and South America, allowing BIOMASS to acquire data over the tropics. Regardless of the frequency transmit permission, however, the spatial resolution of BIOMASS (200 m or coarser) is not as high as desired, as noted by many workshop participants.

Numerous participants said that most of the advances in permafrost mapping in the near future are likely to come from studies based on the use of airborne instrumentation coupled with field measurements used for calibration and validation efforts and associated model development. For example, P-band stripmap SAR data can be acquired across large regions, where military radars and communications are not affected, to retrieve active layer and other subsurface properties. The potential of using multiple-receiver multiple-transmitter airborne radars, at low frequencies (P-band and lower), has also been demonstrated for 3D imaging of subsurface features and should be exploited here as well, said several participants. L-band InSAR data can be acquired over areas where increases in seasonal thaw and long-term permafrost degradation create substantial deformation of the surface because of changes in subsurface moisture/ice content. High-resolution LiDAR data can be acquired over areas where there is thermokarst activity, often indicative of rapid permafrost degradation. AEM can be acquired over areas where boreholes and other field measurements allow the AEM data to be calibrated to map permafrost extent, depth to top of permafrost, permafrost thickness, near-surface soil moisture and ice content, and other variables of interest. Each of these airborne observations allows upscaling of field measurements to much larger areas while building on the information that can be discerned from current and planned spaceborne missions. This multitiered approach, scaling from field to aircraft to satellite observations, would allow high-resolution and spatially extensive retrievals and would thus enable the use of satellite observations over regions where no aircraft or field measurements exist.

As discussed by numerous participants, progress in improving models—both land surface/subsurface permafrost and hydrological models and remote sensing models for developing advanced retrieval algorithms—must be made in the immediate future. In situ and remotely sensed data can be used as model inputs and calibration data sets, and/or for validation. Multisensor data fusion and modeling have been extremely successful in the atmospheric reanalysis community and could be employed to a greater extent in land surface modeling, particularly in permafrost regions. Steps are being taken in this direction by programs such as NASA's Arctic Boreal Vulnerability Experiment, the Department of the Interior's and the U.S. Geological Survey's Integrated Ecosystem Model Project, and the Department of Energy's Next Generation Ecosystem Experiment, all of which could be leveraged in advancing remote sensing of permafrost.

References

Alasset, P. J., V. Poncos, V. Singhroy, and R. Couture. 2008. InSAR Monitoring of Permafrost Activity in the Lower Mackenzie Valley, Canada. Presented at 2008 IEEE International Geoscience and Remote Sensing Symposium.

Anderson, K. and H. Croft. 2009. Remote sensing of soil surface properties. Progress in Physical Geography 33(4):457-473.

Anisimov, O. A., D. G. Vaughan, T. V. Callaghan, C. Furgal, H. Marchant, T. D. Prowse, H. Vilhjálmsson, and J. E. Walsh. 2007. Polar regions. In Climate Change 2007: Impacts, Adaptation and Vulnerability. Contribution of Working Group II to the Fourth Assessment Report of the Intergovernmental Panel on Climate Change. Parry, M. L., O. F. Canziani, J. P. Palutikof, P. J. van der Linden, and C. E. Hanson, eds. Cambridge, UK: Cambridge University Press.

Annan, A. P. and J. L. Davis. 1976. Impulse radar sounding in permafrost. Radio Science 11(4):383-394.

Arcone, S. A., D. E. Lawson, A. J. Delaney, J. C. Strasser, and J. D. Strasser. 1998. Ground-penetrating radar reflection profiling of groundwater and bedrock in an area of discontinuous permafrost. Geophysics 63(5):1573-1584.

Arcone, S. A., M. L. Prentice, and A. J. Delaney. 2002. Stratigraphic profiling with ground-penetrating radar in permafrost: A review of possible analogs for Mars. Journal of Geophysical Research-Planets 107(E11).

Arp, C. D., B. M. Jones, M. Whitman, A. Larsen, and F. E. Urban. 2010. Lake temperature and ice cover regimes in the Alaskan subarctic and arctic: Integrated monitoring, remote sensing, and modeling. Journal of the American Water Resources Association 46(4):777-791.

Arp, C. D., B. M. Jones, F. E. Urban, and G. Grosse. 2011. Hydrogeomorphic processes of thermokarst lakes with grounded-ice and floating-ice regimes on the Arctic coastal plain, Alaska. Hydrological Processes 25(15):2422-2438.

Arp, C. D., B. M. Jones, Z. Lu, and M. S. Whitman. 2012. Shifting balance of thermokarst lake ice regimes across the Arctic Coastal Plain of northern Alaska. Geophysical Research Letters 39.

Arzhanov, M. M., A. V. Eliseev, P. F. Demchenko, I. I. Mokhov, and V. C. Khon. 2008. Simulation of thermal and hydrological regimes of Siberian river watersheds under permafrost conditions from reanalysis data. Izvestiya Atmospheric and Oceanic Physics 44(1):83-89.

Asner, G. P., D. E. Knapp, M. O. Jones, T. Kennedy-Bowdoin, R. E. Martin, J. Boardman, and C. B. Field. 2007. Carnegie Airborne Observatory: In-flight fusion of hyperspectral imaging and waveform light detection and ranging (wLiDAR) for three-dimensional studies of ecosystems. Journal of Applied Remote Sensing 1, doi:10.1117/1.2794018.

Barnes, E. M., K. A. Sudduth, J. W. Hummel, S. M. Lesch, D. L. Corwin, C. H. Yang, C. S. T. Daughtry, and W. C. Bausch. 2003. Remote- and ground-based sensor techniques to map soil properties. Photogrammetric Engineering and Remote Sensing 69(6):619-630.

Bartsch, A., T. Melzer, K. Elger, and B. Heim. 2012. Soil Moisture from METOP ASCAT Data at High Latitudes. Presented at 10th International Conference on Permafrost.

Beck, P. S. A., S. J. Goetz, M. C. Mack, H. D. Alexander, Y. F. Jin, J. T. Randerson, and M. M. Loranty. 2011a. The impacts and implications of an intensifying fire regime on Alaskan boreal forest composition and albedo. Global Change Biology 17(9):2853-2866.

Beck, P. S. A., G. P. Juday, C. Alix, V. A. Barber, S. E. Winslow, E. E. Sousa, P. Heiser, J. D. Herriges, and S. J. Goetz. 2011b. Changes in forest productivity across Alaska consistent with biome shift. Ecology Letters 14(4):373-379.

Belshe, E. F., E. A. G. Schuur, and G. Grosse. 2013. Quantification of upland thermokarst features with high resolution remote sensing. Environmental Research Letters 8(035016):10.

Blackard, J. A., M. V. Finco, E. H. Helmer, G. R. Holden, M. L. Hoppus, D. M. Jacobs, A. J. Lister, G. G. Moisen, M. D. Nelson, R. Riemann, B. Ruefenacht, D. Salajanu, D. L. Weyermann, K. C. Winterberger, T. J. Brandeis, R. L. Czaplewski, R. E. McRoberts, P. L. Patterson, and R. P. Tymcio. 2008. Mapping US forest biomass using nationwide forest inventory data and moderate resolution information. Remote Sensing of Environment 112(4):1658-1677.

Boehnke, K. and V. R. Wismann. 1996. Thawing of Soils in Siberia Observed by the ERS-1 Scatterometer between 1992 and 1995. Presented at 1996 IEEE International Geoscience and Remote Sensing Symposium. Remote Sensing for a Sustainable Future, Lincoln, NE.

Bowen, Z. H. and R. G. Waltermire. 2002. Evaluation of light detection and ranging (LIDAR) for measuring river corridor topography. Journal of the American Water Resources Association 38(1):33-41.

Brosten, T. R., J. H. Bradford, J. P. McNamara, J. P. Zarnetske, W. Bowden, and M. N. Gooseff. 2006. Profiles of temporal thaw depths beneath two Arctic stream types using ground-penetrating radar. Permafrost and Periglacial Processes 17:341-355.

Brown, J. and N. A. Grave. 1979. Physical and thermal disturbance and protection of permafrost. Special Report 79-5. U.S. Army Cold Regions Research and Engineering Laboratory, Hanover, NH.

Browning, D. M. and M. C. Duniway. 2011. Digital soil mapping in the absence of field training data: A case study using terrain attributes and semiautomated soil signature derivation to distinguish ecological potential. Applied and Environmental Soil Science 2011(12).

Bubier, J. L., B. N. Rock, and P. M. Crill. 1997. Spectral reflectance measurements of boreal wetland and forest mosses. Journal of Geophysical Research-Atmospheres 102(D24):29483-29494.

Carter, L. D. and J. P. Galloway. 1985. Engineering-geologic maps of northern Alaska, Harrison Bay quadrangle. U.S. Geological Survey Open File Report 85-256. U.S. Geological Survey, Washington, DC.

Catapano, I., L. Crocco, Y. Krellmann, G. Triltzsch, and F. Soldovieri. 2012. A tomographic approach for helicopter-borne ground penetrating radar imaging. IEEE Geoscience and Remote Sensing Letters 9(3):378-382.

Chen, F. L., H. Lin, W. Zhou, T. H. Hong, and G. Wang. 2013. Surface deformation detected by ALOS PALSAR small baseline SAR interferometry over permafrost environment of Beiluhe section, Tibet Plateau, China. Remote Sensing of Environment 138:10-18.

Cho, H. J., P. Kirui, and H. Natarajan. 2008. Test of multi-spectral vegetation index for floating and canopy-forming submerged vegetation. International Journal of Environmental Research and Public Health 5(5):477-483.

Christensen, T. R., T. R. Johansson, H. J. Akerman, M. Mastepanov, N. Malmer, T. Friborg, P. Crill, and B. H. Svensson. 2004. Thawing sub-Arctic permafrost: Effects on vegetation and methane emissions. Geophysical Research Letters 31(4).

Corwin, D. L. 2008. Past, present, and future trends of soil electrical conductivity measurements using geophysical methods. In Handbook of Agricultural Geophysics. Allred, J. J. D. B. and M. R. Ehsani, eds. Boca Raton, FL: CRC Press.

Crétaux, J.-F., A. V. Kouraev, F. Papa, M. B. Nguyen, A. Cazenave, N. V. Aladin, and I. S. Plotnikov. 2005. Water balance of the Big Aral sea from satellite remote sensing and in situ observations. Great Lakes Research 31:520-534.

De Pascale, G. P., W. H. Pollard, and K. K. Williams. 2008. Geophysical mapping of ground ice using a combination of capacitive coupled resistivity and ground-penetrating radar, Northwest Territories, Canada. Journal of Geophysical Research-Earth Surface 113(F2).

Derksen, C., A. Walker, and B. Goodison. 2003. A comparison of 18 winter seasons of in situ and passive microwave-derived snow water equivalent estimates in Western Canada. Remote Sensing of Environment 88(3):271-282.

DUE Permafrost Project Consortium. 2012. ESA Data User Element (DUE) Permafrost: Circumpolar Remote Sensing Service for Permafrost (Full Product Set) with links to datasets. PANGAEA.

Duguay, C. R., T. Zhang, D. W. Leverington, and V. E. Romanovsky. 2005. Satellite remote sensing of permafrost and seasonally frozen ground. In Remote Sensing in Northern Hydrology: Measuring Environmental Change, Geophysical Monographs Series, vol. 163. Duguay, C. R. and A. Pietroniro, eds. Washington, DC: American Geophysical Union.

Engram, M., K. W. Anthony, F. J. Meyer, and G. Grosse. 2012. Synthetic aperture radar (SAR) backscatter response from methane ebullition bubbles trapped by thermokarst lake ice. Canadian Journal of Remote Sensing 38(6):667-682.

Entekhabi, D., E. G. Njoku, P. E. O'Neill, K. H. Kellogg, W. T. Crow, W. N. Edelstein, J. K. Entin, S. D. Goodman, T. J. Jackson, J. Johnson, J. Kimball, J. R. Piepmeier, R. D. Koster, N. Martin, K. C. McDonald, M. Moghaddam, S. Moran, R. Reichle, J. C. Shi, M. W. Spencer, S. W. Thurman, L. Tsang and J. Van Zyl. 2010. The Soil Moisture Active Passive (SMAP) Mission. Proceedings of the IEEE 98(5):704-716.

Epting, J., D. Verbyla, and B. Sorbel. 2005. Evaluation of remotely sensed indices for assessing burn severity in interior Alaska using Landsat TM and ETM+. Remote Sensing of Environment 96(3-4):328-339.

Farifteh, J., A. Farshad, and R. J. George. 2006. Assessing salt-affected soils using remote sensing, solute modelling, and geophysics. Geoderma 130(3-4):191-206.

Farouki, O. 1981. Thermal Properties of Soils. CRREL Monograph 81-1. U.S. Army Cold Regions Research and Engineering Laboratory, Hanover, NH.

Fitterman, D. V. and M. Deszcz-Pan. 1998. Helicopter EM mapping of saltwater intrusion in Everglades National Park, Florida. Exploration Geophysics 29:240-243.

French, H. M. and Y. Shur. 2010. The principles of cryostratigraphy. Earth-Science Reviews 101(190-206).

Frolking, S., M. W. Palace, D. B. Clark, J. Q. Chambers, H. H. Shugart, and G. C. Hurtt. 2009. Forest disturbance and recovery: A general review in the context of spaceborne remote sensing of impacts on aboveground biomass and canopy structure. Journal of Geophysical Research-Biogeosciences 114.

Gao, J. 2009. Bathymetric mapping by means of remote sensing: Methods, accuracy and limitations. Progress in Physical Geography 33(1):103-116.

Goetz, S. J., M. C. Mack, K. R. Gurney, J. T. Randerson, and R. A. Houghton. 2007. Ecosystem responses to recent climate change and fire disturbance at northern high latitudes: Observations and model results contrasting northern Eurasia and North America. Environmental Research Letters 2(045031):9.

Goetz, S. J., M. Sun, A. Baccini, and P. S. A. Beck. 2010. Synergistic use of space-borne LiDAR and optical imagery for assessing forest disturbance: An Alaska case study. Journal of Geophysical Research: Biogeosciences 115(G00E0).

Goetz, S. J., B. Bond-Lamberty, B. E. Law, J. A. Hicke, C. Huang, R. A. Houghton, S. McNulty, T. O'Halloran, M. Harmon, A. J. H. Meddens, E. M. Pfeifer, D. Mildrexler, and E. S. Kasischke. 2012. Observations and assessment of forest carbon dynamics following disturbance in North America. Journal of Geophysical Research: Biogeosciences 117(G2).

Gomez, C., R. A. V. Rossel, and A. B. McBratney. 2008. Soil organic carbon prediction by hyperspectral remote sensing and field vis-NIR spectroscopy: An Australian case study. Geoderma 146(3-4):403-411.

Grosse, G., L. Schirrmeister, V. V. Kunitsky, and H. W. Hubberten. 2005. The use of CORONA images in remote sensing of periglacial geomorphology: An illustration from the NE Siberian Coast. Permafrost and Periglacial Processes 16(2):163-172.

Grosse, G., L. Schirrmeister, and T. J. Malthus. 2006. Application of Landsat-7 satellite data and a DEM for the quantification of thermokarst-affected terrain types in the periglacial Lena-Anabar coastal lowland. Polar Research 25(1):51-67.

Grosse, G., B. Jones, and C. Arp. 2012. Thermokarst lakes, drainage, and drained basins. In Treatise on Geomorphology. Shroder, J. F., ed. San Diego, CA: Academic Press.

Groves, D. G. and J. A. Francis. 2002a. Moisture budget of the Arctic atmosphere from TOVS satellite data. Journal of Geophysical Research-Atmospheres 107(D19).

Groves, D. G. and J. A. Francis. 2002b. Variability of the Arctic atmospheric moisture budget from TOVS satellite data. Journal of Geophysical Research-Atmospheres 107(D24).

Hachem, S., M. Allard, and C. Duguay. 2009. Using the MODIS land surface temperature product for mapping permafrost: An Application to Northern Quebec and Labrador, Canada. Permafrost and Periglacial Processes 20(4):407-416.

Hachem, S., C. R. Duguay, and M. Allard. 2012. Comparison of MODIS-derived land surface temperatures with ground surface and air temperature measurements in continuous permafrost terrain. Cryosphere 6(1):51-69.

Hall, D. K. 1982. A review of the utility of remote sensing in Alaskan permafrost studies. IEEE Transactions on Geoscience and Remote Sensing GE-20(390-394).

Hall, D. K. and G. A. Riggs. 2007. Accuracy assessment of the MODIS snow-cover products. Hydrological Processes 21:1534-1547.

Hancock, S., R. Baxter, J. Evans, and B. Huntley. 2013. Evaluating global snow water equivalent products for testing land surface models. Remote Sensing of Environment 128:107-117.

Harden, J. W., K. L. Manies, M. R. Turetsky, and J. C. Neff. 2006. Effects of wildfire and permafrost on soil organic matter and soil climate in interior Alaska. Global Change Biology 12(12):2391-2403.

Hinkel, K. M. and F. E. Nelson. 2003. Spatial and temporal patterns of active layer thickness at Circumpolar Active Layer Monitoring (CALM) sites in northern Alaska, 1995-2000. Journal of Geophysical Research-Atmospheres 108(D2).

Hinkel, K. M., B. M. Jones, W. R. Eisner, C. J. Cuomo, R. A. Beck, and R. Frohn. 2007. Methods to assess natural and anthropogenic thaw lake drainage on the western Arctic coastal plain of northern Alaska. Journal of Geophysical Research-Earth Surface 112(F2).

Hinzman, L. D., N. D. Bettez, W. R. Bolton, F. S. Chapin, M. B. Dyurgerov, C. L. Fastie, B. Griffith, R. D. Hollister, A. Hope, H. P. Huntington, A. M. Jensen, G. J. Jia, T. Jorgenson, D. L. Kane, D. R. Klein, G. Kofinas, A. H. Lynch, A. H. Lloyd, A. D. McGuire, F. E. Nelson, W. C. Oechel, T. E. Osterkamp, C. H. Racine, V. E. Romanovsky, R. S. Stone, D. A. Stow, M. Sturm, C. E. Tweedie, G. L. Vourlitis, M. D. Walker, D. A. Walker, P. J. Webber, J. M. Welker, K. Winker, and K. Yoshikawa. 2005. Evidence and implications of recent climate change in northern Alaska and other arctic regions. Climatic Change 72(3):251-298.

Hively, W. D., G. W. McCarty, J. B. Reeves III, M. W. Lang, R. A. Oesterling, and S. R. Delwiche. 2011. Use of airborne hyperspectral imagery to map soil properties in tilled agricultural fields. Applied and Environmental Soil Science 2011(13).

Ho, L. T. K., Y. Yamaguchi, and M. Umitsu. 2012. Rule-based landform classification by combining multi-spectral/temporal satellite data and the SRTM DEM. International Journal of Geoinformatics 8(27-38).

Hoy, E. E., N. H. F. French, M. R. Turetsky, S. N. Trigg, and E. S. Kasischke. 2008. Evaluating the potential of Landsat TM/ETM+ imagery for assessing fire severity in Alaskan black spruce forests. International Journal of Wildland Fire 17(4):500-514.

Hubbard, S. S., C. Gangodagamage, B. Dafflon, H. Wainwright, J. Peterson, A. Gusmeroli, C. Ulrich, Y. Wu, C. Wilson, J. Rowland, C. Tweedie, and S. D. Wullschleger. 2013. Quantifying and relating land-surface and subsurface variability in permafrost environments using LiDAR and surface geophysical datasets. Hydrogeology Journal 21(1):149-169.

IGOS. 2007. Cryosphere Theme Report. WMO/TD-No. 1405. World Meteorological Organization, Geneva, Switzerland.

Instanes, A. and O. Anisimov. 2008. Climate Change and Arctic infrastructure. Presented at Ninth International Conference on Permafrost, Fairbanks, AK.

Jafarov, E. E., S. S. Marchenko, and V. E. Romanovsky. 2012. Numerical modeling of permafrost dynamics in Alaska using a high spatial resolution dataset. Cryosphere 6(3):613-624.

Jarmer, T., P. Rosso, and M. Ehlers. 2010. Mapping Topsoil Organic Carbon of Agricultural Soils from Hyperspectral Remote Sensing Data. Presented at Hyperspectral 2010 Workshop, March 17-19, 2010, Frascati, Italy.

Jepsen, S. M., C. I. Voss, M. A. Walvoord, B. J. Minsley, and J. Rover. 2013. Linkages between lake shrinkage/expansion and sublacustrine permafrost distribution determined from remote sensing of interior Alaska, USA. Geophysical Research Letters 40(5):882-887.

Jezek, K., X. Wu, P. Gogineni, E. Rodríguez, A. Freeman, F. Rodriguez-Morales, and C. D. Clark. 2011. Radar images of the bed of the Greenland Ice Sheet. Geophysical Research Letters 38(1).

Ji, L., B. K. Wylie, D. R. Nossov, B. Peterson, M. P. Waldrop, J. W. McFarland, J. Rover, and T. N. Hollingsworth. 2012. Estimating aboveground biomass in interior Alaska with Landsat data and field measurements. International Journal of Applied Earth Observation and Geoinformation 18:451-461.

Johnson, K. D., J. W. Harden, A. D. McGuire, M. Clark, F. M. Yuan, and A. O. Finley. 2013. Permafrost and organic layer interactions over a climate gradient in a discontinuous permafrost zone. Environmental Research Letters 8(3).

Jones, A., L. Montanarella, V. Stolobovoy, G. Broll, C. Tarnocai, O. Spaargaren, and C.-L. Ping. 2010. Soil Atlas of the Northern Circumpolar Region. Luxemborg: Publications Office of the European Union.

Jones, B. M., G. Grosse, C. D. Arp, M. C. Jones, K. M. W. Anthony, and V. E. Romanovsky. 2011. Modern thermokarst lake dynamics in the continuous permafrost zone, northern Seward Peninsula, Alaska. Journal of Geophysical Research-Biogeosciences 116.

Jones, B. M., G. Grosse, K. M. Hinkel, C. D. Arp, S. Walker, R. A. Beck, and J. P. Galloway. 2012. Assessment of pingo distribution and morphometry using an IfSAR derived digital surface model, western Arctic Coastal Plain, Northern Alaska. Geomorphology 138(1):1-14.

Jones, B. M., J. M. Stoker, A. E. Gibbs, G. Grosse, V. E. Romanovsky, T. A. Douglas, N. E. M. Kinsman, and B. M. Richmond. 2013. Quantifying landscape change in an Arctic coastal lowland using repeat airborne LiDAR. Environmental Research Letters 8(4).

Jones, H. G. and R. A. Vaughan. 2010. Remote sensing of vegetation: Principles, techniques, and applications. Oxford, New York: Oxford University Press.

Jorgenson, M. T., Y. Shur, and H. J. Walker. 1998. Evolution of a permafrost dominated landscape on the Colville River Delta, northern Alaska. Collection Nordicana 55:523-530.

Jorgenson, M. T., C. H. Racine, J. C. Walters, and T. E. Osterkamp. 2001. Permafrost degradation and ecological changes associated with a warming climate in central Alaska. Climatic Change 48(4):551-579.

Jorgenson, M. T., K. Yoshikawa, M. Kanveskiy, Y. L. Shur, V. Romanovsky, S. Marchenko, G. Grosse, J. Brown, and B. Jones. 2008. Permafrost Characteristics of Alaska. In Proceedings of the Ninth International Conference on Permafrost. Kane, D. L. and K. M. Hinkel, eds. Fairbanks: University of Alaska, Fairbanks.

Jorgenson, M. T., J. E. Roth, P. F. Miller, M. J. Macander, M. S. Duffy, A. F. Wells, G. V. Frost, and E. R. Pullman. 2009. An Ecological Land Survey and Landcover Map of the Arctic Network. Natural Resource Technical Report NPS/ARCN/NRTR—2009/270. National Park Service, Fort Collins, CO.

Jorgenson, M. T., V. Romanovsky, J. Harden, Y. Shur, J. O'Donnell, E. A. G. Schuur, M. Kanevskiy, and S. Marchenko. 2010. Resilience and vulnerability of permafrost to climate change. Canadian Journal of Forest Research-Revue Canadienne De Recherche Forestiere 40(7):1219-1236.

Jorgenson, M. T., M. Kanevskiy, Y. Shur, T. Osterkamp, D. Fortier, T. Cater, and P. Miller. 2012. Thermokarst Lake and Shore Fen Development in Boreal Alaska. Presented at Tenth International Conference on Permafrost, June 25-29, 2012, Salekhard, Russia.

Jorgenson, M. T., J. Harden, M. Kanevskiy, J. O'Donnell, K. Wickland, S. Ewing, K. Manies, Q. L. Zhuang, Y. Shur, R. Striegl, and J. Koch. 2013. Reorganization of vegetation, hydrology and soil carbon after permafrost degradation across heterogeneous boreal landscapes. Environmental Research Letters 8(3).

Kääb, A. 2008. Remote sensing of permafrost-related problems and hazards. Permafrost and Periglacial Processes 19(2):107-136.

Kanevskiy, M., Y. Shur, D. Fortier, M. T. Jorgenson, and E. Stephani. 2011. Cryostratigraphy of late Pleistocene syngenetic permafrost (yedoma) in northern Alaska, Itkillik River exposure. Quaternary Research 75(3):584-596.

Kanevskiy, M., Y. Shur, M. T. Jorgenson, C. L. Ping, G. J. Michaelson, D. Fortier, E. Stephani, M. Dillon, and V. Tumskoy. 2013. Ground ice in the upper permafrost of the Beaufort Sea coast of Alaska. Cold Regions Science and Technology 85:56-70.

Kasischke, E. S., D. L. Verbyla, T. S. Rupp, A. D. McGuire, K. A. Murphy, R. Jandt, J. L. Barnes, E. E. Hoy, P. A. Duffy, M. Calef, and M. R. Turetsky. 2010. Alaska's changing fire regime—implications for the vulnerability of its boreal forests. Canadian Journal of Forest Research-Revue Canadienne De Recherche Forestiere 40(7):1313-1324.

Kelly, R. E., A. T. Chang, L. Tsang, and J. L. Foster. 2003. A prototype AMSR-E global snow area and snow depth algorithm. IEEE Transactions on Geoscience and Remote Sensing 41(2):230-242.

Kheyrollah Pour, H. K., C. R. Duguay, A. Martynov, and L. C. Brown. 2012. Simulation of surface temperature and ice cover of large northern lakes with 1-D models: A comparison with MODIS satellite data and in situ measurements. Tellus Series a-Dynamic Meteorology and Oceanography 64.

Kimball, J. S., K. C. McDonald, and M. Zhao. 2006. Spring thaw and its effect on terrestrial vegetation productivity in the western Arctic observed from satellite microwave and optical remote Sensing. Earth Interactions 10:1-22.

Kimball, J. S., L. A. Jones, K. Zhang, F. A. Heinsch, K. C. McDonald, and W. C. Oechel. 2009. A satellite approach to estimate land-atmosphere CO_2 exchange for boreal and Arctic biomes using MODIS and AMSR-E. IEEE Transactions on Geoscience and Remote Sensing 47(2):569-587.

Komarov, S. A., V. L. Mironov, and S. Li. 2002. SAR polarimetry for permafrost active layer freeze/thaw processes. Presented at 2002 IEEE International Geoscience and Remote Sensing Symposium.

Kravtsova, V. I. and A. G. Bystrova. 2009. Changes in thermokarst lake sizes in different regions of Russia for the last 30 years. Kriosfera Zemli (Earth Cryosphere) 13:16-26.

Kreig, R. A. 1977. Terrain analysis for the Trans-Alaska pipeline. Civil Engineering ASCE 47:61-65.

Kreig, R. A. and R. D. Reger. 1982. Air-photo analysis and summary of landform soil properties along the route of the Trans-Alaska Pipeline System. Geologic Report 66. Alaska Division of Geological and Geophysical Surveys, Fairbanks, AK.

Laamrani, A., O. Valeria, L. Z. Cheng, Y. Bergeron, and C. Camerlynck. 2013. The use of ground penetrating radar for remote sensing the organic layer-mineral soil interface in paludified boreal forests. Canadian Journal of Remote Sensing 39:74-88.

Labrecque, S., D. Lacelle, C. R. Duguay, B. Lauriol, and J. Hawkings. 2009. Contemporary (1951-2001) evolution of lakes in the Old Crow Basin, Northern Yukon, Canada: Remote sensing, numerical modeling, and stable isotope analysis. Arctic 62(2):225-238.

Langer, M., S. Westermann, M. Heikenfeld, W. Dorn, and J. Boike. 2013. Satellite-based modeling of permafrost temperatures in a tundra lowland landscape. Remote Sensing of Environment 135:12-24.

Larsen, P. H., S. Goldsmith, O. Smith, M. L. Wilson, K. Strzepek, P. Chinowsky, and B. Saylor. 2008. Estimating future costs for Alaska public infrastructure at risk from climate change. Global Environmental Change-Human and Policy Dimensions 18(3):442-457.

Lee, H., M. Durand, H. C. Jung, D. Alsdorf, C. K. Shum, and Y. W. Sheng. 2010. Characterization of surface water storage changes in Arctic lakes using simulated SWOT measurements. International Journal of Remote Sensing 31(14):3931-3953.

Leuschen, C., P. Kanagaratnam, K. Yoshikawa, S. Arcone, and S. Gogineni. 2003. Design and field experiments of a surface-penetrating radar for Mars exploration. Journal of Geophysical Research 108(E4):8034.

Liu, L., T. J. Zhang, and J. Wahr. 2010. InSAR measurements of surface deformation over permafrost on the North Slope of Alaska. Journal of Geophysical Research-Earth Surface 115.

Liu, L., K. Schaefer, T. J. Zhang, and J. Wahr. 2012. Estimating 1992-2000 average active layer thickness on the Alaskan North Slope from remotely sensed surface subsidence. Journal of Geophysical Research-Earth Surface 117.

Liu, Q., R. H. Reichle, R. Bindlish, M. H. Cosh, W. T. Crow, R. de Jeu, G. J. M. De Lannoy, G. J. Huffman, and T. J. Jackson. 2011. The contributions of precipitation and soil moisture observations to the skill of soil moisture estimates in a land data assimilation system. Journal of Hydrometeorology 12(5):750-765.

Liu, Y. H., J. R. Key, A. Schweiger, and J. Francis. 2006. Characteristics of satellite-derived clear-sky atmospheric temperature inversion strength in the Arctic, 1980 96. Journal of Climate 19(19):4902-4913.

Loisel, J., Z. C. Yu, A. Parsekian, J. Nolan, and L. Slater. 2013. Quantifying landscape morphology influence on peatland lateral expansion using ground-penetrating radar (GPR) and peat core analysis. Journal of Geophysical Research-Biogeosciences 118(2):373-384.

Lucieer, A., S. Robinson, D. Turner, S. Harwin, and J. Kelcey. 2012. Using a micro-UAV for ultra-high resolution multi-sensor observations of Antarctic moss beds. International Archives of the Photogrammetry, Remote Sensing and Spatial Information Sciences XXXIX-B1:429-433.

Marchenko, S., V. Romanovsky, and G. Tipenko. 2008. Numerical Modeling of Spatial Permafrost Dynamics in Alaska. Presented at Ninth International Conference on Permafrost.

Marchenko, S., S. Hachem, V. Romanovsky, and C. Duguay. 2009. Permafrost and Active Layer Modeling in the Northern Eurasia Using MODIS Land Surface Temperature as an Input Data. Presented at EGU General Assembly 2009, Vienna, Austria, 19-24 April, 2009.

Matti Vaaja, M. and M. Hallikainen. 2013. Remote Sensing of Frozen Soil at UHF Frequencies. Presented at XXXIII Finnish URSI Convention on Radio Science and SMARAD Seminar.

McDonald, K. C., J. S. Kimball, E. Njoku, R. Zimmermann, and M. S. Zhao. 2004. Variability in springtime thaw in the terrestrial high latitudes: Monitoring a major control on the biospheric assimilation of atmospheric CO(2) with spaceborne microwave remote sensing. Earth Interactions 8.

McGuire, A. D., C. Wirth, M. Apps, J. Beringer, J. Clein, H. Epstein, D. W. Kicklighter, J. Bhatti, F. S. Chapin, B. de Groot, D. Efremov, W. Eugster, M. Fukuda, T. Gower, L. Hinzman, B. Huntley, G. J. Jia, E. Kasischke, J. Melillo, V. Romanovsky, A. Shvidenko, E. Vaganov, and D. Walker. 2002. Environmental variation, vegetation distribution, carbon dynamics and water/energy exchange at high latitudes. Journal of Vegetation Science 13(3):301-314.

McGuire, A. D., L. G. Anderson, T. R. Christensen, S. Dallimore, L. D. Guo, D. J. Hayes, M. Heimann, T. D. Lorenson, R. W. Macdonald, and N. Roulet. 2009. Sensitivity of the carbon cycle in the Arctic to climate change. Ecological Monographs 79(4):523-555.

Minsley, B. J., J. D. Abraham, B. D. Smith, J. C. Cannia, C. I. Voss, M. T. Jorgenson, M. A. Walvoord, B. K. Wylie, L. Anderson, L. B. Ball, M. Deszcz-Pan, T. P. Wellman, and T. A. Ager. 2012. Airborne electromagnetic imaging of discontinuous permafrost. Geophysical Research Letters 39.

Mishra, U. 2013. Soil health and climate change. Soil Science Society of America Journal 77(1):336.

Mishra, U. and W. J. Riley. 2012. Alaskan soil carbon stocks: Spatial variability and dependence on environmental factors. Biogeosciences 9(9):3637-3645.

Moghaddam, M., S. Saatchi, and R. H. Cuenca. 2000. Estimating subcanopy soil moisture with radar. Journal of Geophysical Research-Atmospheres 105(D11):14899-14911.

Moorman, B. J., S. D. Robinson, and M. M. Burgess. 2003. Imaging periglacial conditions with ground-penetrating radar. Permafrost and Periglacial Processes 14(4):319-329.

Morris, D. K., K. W. Ross, and C. J. Johannsen. 2008. The characterization of soil properties to develop soil management/mapping units using high-resolution remotely sensed data sets. The Journal of Terrestrial Observation 1:5-37.

Mugford, R. I. and J. A. Dowdeswell. 2010. Modeling iceberg-rafted sedimentation in high-latitude fjord environments. Journal of Geophysical Research-Earth Surface 115.

Mulder, V. L., S. de Bruin, M. E. Schaepman, and T. R. Mayr. 2011. The use of remote sensing in soil and terrain mapping—A review. Geoderma 162(1-2):1-19.

Murton, J. B. 2013. Ground ice and cryostratigraphy. In Treatise on Geomorphology. Shroder, J. F., ed. San Diego: Academic Press.

Muskett, R. R. and V. E. Romanovsky. 2009. Groundwater storage changes in Arctic permafrost watersheds from GRACE and in situ measurements. Environmental Research Letters 4(4).

Muskett, R. R. and V. E. Romanovsky. 2011. Alaskan permafrost groundwater storage changes derived from GRACE and ground measurements. Remote Sensing 3(2):378-397.

Neigh, C. S. R., R. F. Nelson, K. J. Ranson, H. A. Margolis, P. M. Montesano, G. Q. Sun, V. Kharuk, E. Naesset, M. A. Wulder, and H. E. Andersen. 2013. Taking stock of circumboreal forest carbon with ground measurements, airborne and spaceborne LiDAR. Remote Sensing of Environment 137:274-287.

Nossov, D. R., M. T. Jorgenson, K. Kielland, and M. Z. Kanevskiy. 2013. Edaphic and microclimatic controls over permafrost response to fire in interior Alaska. Environmental Research Letters 8(3).

NRC (National Research Council). 2007. Earth Science and Applications from Space: National Imperatives for the Next Decade and Beyond. Washington, DC: The National Academies Press.

NRC. 2012. Earth Science and Applications from Space: A Midterm Assessment of NASA's Implementation of the Decadal Survey. Washington, DC: The National Academies Press.

Osterkamp, T. E., M. T. Jorgenson, E. A. G. Schuur, Y. L. Shur, M. Z. Kanevskiy, J. G. Vogel, and V. E. Tumskoy. 2009. Physical and ecological changes associated with warming permafrost and thermokarst in interior Alaska. Permafrost and Periglacial Processes 20(3):235-256.

Paden, J., T. Akins, D. Dunson, C. Allen, and P. Gogineni. 2010. Ice-sheet bed 3-D tomography. Journal of Glaciology 56(195).

Paine, J. G. and B. R. S. Minty. 2005. Airborne hydrogeophysics. In Hydrogeophysics. Rubin, Y. and S. S. Hubbard, eds. Dordrecht, Netherlands: Springer.

Paine, J. G., J. R. Andrews, K. Saylam, T. A. Tremblay, A. R. Averett, T. L. Caudle, T. Meyer, and M. H. Young. 2013. Airborne lidar on the Alaskan North Slope: Wetlands mapping, lake volumes, and permafrost features. The Leading Edge 32(7):798-805.

Panda, S. K., A. Prakash, D. N. Solie, V. E. Romanovsky, and M. T. Jorgenson. 2010. Remote sensing and field-based mapping of permafrost distribution along the Alaska Highway corridor, Interior Alaska. Periglacial Processes 21(271-281).

Panda, S. K., A. Prakash, M. T. Jorgenson, and D. N. Solie. 2012. Near-surface permafrost distribution mapping using logistic regression and remote sensing in Interior Alaska. GISciences & Remote Sensing 49(3):346-363.

Park, S.-E., A. Bartsch, D. Sabel, and W. Wagner. 2010. Monitoring of Thawing Process Using ENVISAT ASAR Global Mode Data. Presented at Geoscience and Remote Sensing Symposium (IGARSS), 2010 IEEE International.

Parsekian, A. D., B. M. Jones, M. Jones, G. Grosse, K. M. W. Anthony, and L. Slater. 2011. Expansion rate and geometry of floating vegetation mats on the margins of thermokarst lakes, northern Seward Peninsula, Alaska, USA. Earth Surface Processes and Landforms 36(14):1889-1897.

Pastick, N. J., M. T. Jorgenson, B. K. Wylie, B. J. Minsley, L. Ji, M. A. Walvoord, B. D. Smith, J. D. Abraham, and J. R. Rose. 2013. Extending airborne electromagnetic surveys for regional active layer and permafrost mapping with remote sensing and ancillary data, Yukon Flats Ecoregion, Central Alaska. Permafrost and Periglacial Processes 24(3):184-199.

Peddle, D. R. and S. E. Franklin. 1993. Classification of permafrost active layer depth from remotely sensed and topographic evidence. Remote Sensing of Environment 44(1):67-80.

Ping, C. L., G. J. Michaelson, L. D. Guo, M. T. Jorgenson, M. Kanevskiy, Y. Shur, F. G. Dou, and J. J. Liang. 2011. Soil carbon and material fluxes across the eroding Alaska Beaufort Sea coastline. Journal of Geophysical Research: Biogeosciences 116.

Rawlinson, S. E. 1993. Surficial geology and morphology of the Alaskan Central Arctic Coastal Plain. Fairbanks: State of Alaska, Department of Natural Resources, Division of Geological & Geophysical Surveys.

Raynolds, M. K., D. A. Walker, H. E. Epstein, J. E. Pinzon, and C. J. Tucker. 2012. A new estimate of tundra-biome phytomass from trans-Arctic field data and AVHRR NDVI. Remote Sensing Letters 3(5):403-411.

Revercomb, H. E., D. C. Tobin, R. O. Knuteson, J. K. Taylor, D. Deslover, L. Borg, G. Martin, and G. Quinn. 2013. Suomi NPP/JPSS Cross-track Infrared Sounder (CrIS): Radiometric and Spectral Performance. Presented at 93rd American Meteorological Society Annual Meeting, Austin, TX.

Riseborough, D., N. Shiklomanov, B. Etzelmuller, S. Gruber, and S. Marchenko. 2008. Recent advances in permafrost modelling. Permafrost and Periglacial Processes 19(2):137-156.

Rodriguez-Morales, F., P. Gogineni, C. Leuschen, J. Paden, J. Li, C. Lewis, B. Panzer, D. G.-G. Alvestegui, A. Patel, K. Byers, R. Crowe, K. Player, R. Hale, E. Arnold, L. Smith, C. Gifford, D. Braaten, and C. Panton. 2013. Advanced multifrequency radar instrumentation for polar research. IEEE Transactions on Geoscience and Remote Sensing PP(99):1-19.

Romanovsky, V. E., S. L. Smith, and H. H. Christiansen. 2010. Permafrost thermal state in the polar Northern Hemisphere during the International Polar Year 2007-2009: A synthesis. Permafrost and Periglacial Processes 21(2):106-116.

Rosa, E., M. Larocque, S. Pellerin, S. Gagne, and R. Fournier. 2009. Determining the number of manual measurements required to improve peat thickness estimations by ground penetrating radar. Earth Surface Processes and Landforms 34(3):377-383.

Rott, H., S. H. Yueh, D. W. Cline, C. Duguay, R. Essery, C. Haas, F. Heliere, M. Kern, G. Macelloni, E. Malnes, T. Nagler, J. Pulliainen, H. Rebhan, and A. Thompson. 2010. Cold regions hydrology high-resolution observatory for snow and cold land processes. Proceedings of the IEEE 98(5):752-765.

Rott, H., C. Duguay, P. Etchevers, R. Essery, I. Hajnsek, G. Macelloni, E. Malnes, and J. Pulliainen. 2012. Report for Mission Selection: CoReH2O. ESA SP-1324/2. European Space Agency, Noordwijk, The Netherlands.

Rowland, J. C., C. E. Jones, G. Altmann, R. Bryan, B. T. Crosby, L. D. Hinzman, D. L. Kane, D. M. Lawrence, A. Mancino, P. Marsh, J. P. McNamara, V. E. Romanvosky, H. Toniolo, B. J. Travis, E. Trochim, C. J. Wilson, and G. L. Geernaert. 2010. Arctic Landscapes in Transition: Responses to Thawing Permafrost. Eos, Transactions American Geophysical Union 91(26):229-230. DOI: 10.1029/2010EO260001.

Sabel, D., A. Bartsch, S. Schlaffer, J. P. Klein, and W. Wagner. 2012. Soil moisture mapping in permafrost regions—An outlook to Sentinel-1. Presented at 2012 IEEE International Geoscience and Remote Sensing Symposium (IGARSS).

Sannel, A. B. K. and I. A. Brown. 2010. High-resolution remote sensing identification of thermokarst lake dynamics in a subarctic peat plateau complex. Canadian Journal of Remote Sensing 36:S26-S40.

Sazonova, T. S. and V. E. Romanovsky. 2003. A model for regional-scale estimation of temporal and spatial variability of active layer thickness and mean annual ground temperatures. Permafrost and Periglacial Processes 14(2):125-139.

Sazonova, T. S., V. E. Romanovsky, J. E. Walsh, and D. O. Sergueev. 2004. Permafrost dynamics in the 20th and 21st centuries along the East Siberian transect. Journal of Geophysical Research-Atmospheres 109(D1).

Schuur, E. A. G., J. Bockheim, J. G. Canadell, E. Euskirchen, C. B. Field, S. V. Goryachkin, S. Hagemann, P. Kuhry, P. M. Lafleur, H. Lee, G. Mazhitova, F. E. Nelson, A. Rinke, V. E. Romanovsky, N. Shiklomanov, C. Tarnocai, S. Venevsky, J. G. Vogel, and S. A. Zimov. 2008. Vulnerability of permafrost carbon to climate change: Implications for the global carbon cycle. Bioscience 58(8):701-714.

Selkowitz, D. J. and S. V. Stehman. 2011. Thematic accuracy of the National Land Cover Database (NLCD) 2001 land cover for Alaska. Remote Sensing of Environment 115(6):1401-1407.

Selkowitz, D. J., G. Green, B. Peterson, and B. Wylie. 2012. A multi-sensor lidar, multi-spectral and multi-angular approach for mapping canopy height in boreal forest regions. Remote Sensing of Environment 121:458-471.

Short, N., B. Brisco, N. Couture, W. Pollard, K. Murnaghan, and P. Budkewitsch. 2011. A comparison of TerraSAR-X, RADARSAT-2 and ALOS-PALSAR interferometry for monitoring permafrost environments, case study from Herschel Island, Canada. Remote Sensing of Environment 115(12):3491-3506.

Shur, Y. L. and M. T. Jorgenson. 2007. Patterns of permafrost formation and degradation in relation to climate and ecosystems. Permafrost and Periglacial Processes 18(1):7-19.

Siemon, B. 2006. Electromagnetic methods—frequency domain. In Groundwater Geophysics: A Tool for Hydrogeology. Kirsch, R., ed. Berlin: Springer-Verlag.

Singhroy, V., R. Couture, P. J. Alasset, and V. Poncos. 2007. InSAR monitoring of landslides on permafrost terrain in Canada. Presented at IGARSS 2007 (Geoscience and Remote Sensing Symposium, 2007).

Slater, L. D. and A. Reeve. 2002. Investigating peatland stratigraphy and hydrogeology using integrated electrical geophysics. Geophysics 67(2):365-378.

Smith, S. L., V. E. Romanovsky, A. G. Lewkowicz, C. R. Burn, M. Allard, G. D. Clow, K. Yoshikawa, and J. Throop. 2010. Thermal state of permafrost in North America: A contribution to the International Polar Year. Permafrost and Periglacial Processes 21(2):117-135.

Solberg, S., R. Astrup, J. Breidenbach, B. Nilsen, and D. Weydahl. 2013. Monitoring spruce volume and biomass with InSAR data from TanDEM-X. Remote Sensing of Environment 139:60-67.

Stow, D., S. Daeschner, A. Hope, D. Douglas, A. Petersen, R. Myneni, L. Zhou, and W. Oechel. 2003. Variability of the seasonally integrated normalized difference vegetation index across the north slope of Alaska in the 1990s. International Journal of Remote Sensing 24(5):1111-1117.

Stow, D. A., A. Hope, D. McGuire, D. Verbyla, J. Gamon, F. Huemmrich, S. Houston, C. Racine, M. Sturm, K. Tape, L. Hinzman, K. Yoshikawa, C. Tweedie, B. Noyle, C. Silapaswan, D. Douglas, B. Griffith, G. Jia, H. Epstein, D. Walker, S. Daeschner, A. Petersen, L. M. Zhou, and R. Myneni. 2004. Remote sensing of vegetation and land-cover change in Arctic Tundra Ecosystems. Remote Sensing of Environment 89(3):281-308.

Strozzi, T., R. Delaloye, A. Kaab, C. Ambrosi, E. Perruchoud, and U. Wegmuller. 2010. Combined observations of rock mass movements using satellite SAR interferometry, differential GPS, airborne digital photogrammetry, and airborne photography interpretation. Journal of Geophysical Research-Earth Surface 115.

Surdu, C., C. R. Duguay, L. C. Brown, and D. F. Prieto. 2013. Response of ice cover on shallow Arctic lakes of the North Slope of Alaska to contemporary climate conditions (1950-2011): Radar remote sensing and numerical modeling data analysis. The Cryosphere Discussion 7:3783-3821.

Tabatabaeenejad, A. and M. Moghaddam. 2011. Radar retrieval of surface and deep soil moisture and effect of moisture profile on inversion accuracy. IEEE Geoscience and Remote Sensing Letters 8(3):478-482.

Tabatabaeenejad, A., M. Burgin, and M. Moghaddam. In review. P-band radar retrieval of subcanopy and subsurface soil moisture profile as a second-order polynomial: First AirMOSS results. IEEE Transactions on Geoscience and Remote Sensing.

Tarnocai, C., J. G. Canadell, E. A. G. Schuur, P. Kuhry, G. Mazhitova, and S. Zimov. 2009. Soil organic carbon pools in the northern circumpolar permafrost region. Global Biogeochemical Cycles 23.

Thurner, M., C. Beer, M. Santoro, N. Carvalhais, T. Wutzler, D. Schepaschenko, A. Shvidenko, E. Kompter, B. Ahrens, S. R. Levick, and C. Schmullius. 2013. Carbon stock and density of northern boreal and temperate forests. Global Ecology and Biogeography.

Trucco, C., E. A. G. Schuur, S. M. Natali, E. F. Belshe, R. Bracho, and J. Vogel. 2012. Seven-year trends of CO_2 exchange in a tundra ecosystem affected by long-term permafrost thaw. Journal of Geophysical Research: Biogeosciences 117.

Ulaby, F., D. Long, W. Blackwell, C. Elachi, A. Fung, C. Ruf, K. Sarabandi, J. v. Zyl, and H. Zebker. 2013. Microwave Radar and Radiometric Remote Sensing. Ann Arbor, MI: University of Michigan Press.

Van Dam, R. L. 2012. Landform characterization using geophysics—Recent advances, applications, and emerging tools. Geomorphology 137(1):57-73.

van Everdingen, R., ed. 2005. Multi-language glossary of permafrost and related ground-ice terms. Boulder, CO: National Snow and Ice Data Center/World Data Center for Glaciology.

Velicogna, I., J. Tong, T. Zhang, and J. S. Kimball. 2012. Increasing subsurface water storage in discontinuous permafrost areas of the Lena River basin, Eurasia, detected from GRACE. Geophysical Research Letters 39.

Verbyla, D. 2008. The greening and browning of Alaska based on 1982-2003 satellite data. Global Ecology and Biogeography 17(4):547-555.

Vourlitis, G. L., W. C. Oechel, S. J. Hastings, and M. A. Jenkins. 1993. The effect of soil-moisture and thaw depth on CH4 flux from wet coastal tundra ecosystems on the north slope of Alaska. Chemosphere 26(1-4):329-337.

Walter, K. M., S. A. Zimov, J. P. Chanton, D. Verbyla, and F. S. Chapin. 2006. Methane bubbling from Siberian thaw lakes as a positive feedback to climate warming. Nature 443(7107):71-75.

Walter, K. M., M. E. Edwards, G. Grosse, S. A. Zimov, and F. S. Chapin. 2007a. Thermokarst lakes as a source of atmospheric CH4 during the last deglaciation. Science 318(5850):633-636.

Walter, K. M., L. C. Smith, and F. S. Chapin. 2007b. Methane bubbling from northern lakes: Present and future contributions to the global methane budget. Philosophical Transactions of the Royal Society A-Mathematical Physical and Engineering Sciences 365(1856):1657-1676.

Walter, K. M., M. Engram, C. R. Duguay, M. O. Jeffries, and F. S. Chapin. 2008. The potential use of synthetic aperture radar for estimating methane ebullition from Arctic lakes. Journal of the American Water Research Association 44(2):305-315.

Watanabe, M., G. Kadosaki, Y. Kim, M. Ishikawa, K. Kushida, Y. Sawada, T. Tadono, M. Fukuda, and M. Sato. 2012. Analysis of the sources of variation in L-band backscatter from terrains with permafrost. IEEE Transactions on Geoscience and Remote Sensing 50(1):44-54.

Weng, Y.-L., P. Gong, and Z.-L. Zhu. 2010. A spectral index for estimating soil salinity in the Yellow River Delta Region of China using EO-1 Hyperion data. Pedosphere 20(378-388).

Westermann, S., U. Wollschlager, and J. Boike. 2010. Monitoring of active layer dynamics at a permafrost site on Svalbard using multi-channel ground-penetrating radar. Cryosphere 4(4):475-487.

Westermann, S., C. R. Duguay, G. Grosse, and A. Kääb. In press. Remote sensing of permafrost and frozen ground. In Remote Sensing of the Cryosphere. Tedesco, M., ed. Oxford, UK: Wiley-Blackwell.

Whitcomb, J., M. Moghaddam, K. McDonald, J. Kellndorfer, and E. Podest. 2009. Mapping vegetated wetlands of Alaska using L-band radar satellite imagery. Canadian Journal of Remote Sensing 35(1):54-72.

WMO (World Meteorological Society). 2010. GCOS—Global Climate Observing System. http://gcos.wmo.int.

Wulder, M. A., T. Han, J. C. White, T. Sweda, and H. Tsuzuki. 2007. Integrating profiling LIDAR with Landsat data for regional boreal forest canopy attribute estimation and change characterization. Remote Sensing of Environment 110(1):123-137.

Xu, X. L., L. Tsang, and S. Yueh. 2012. Electromagnetic models of co/cross polarization of bicontinuous/DMRT in radar remote sensing of terrestrial snow at X- and Ku-band for CoReH2O and SCLP applications. IEEE Journal of Selected Topics in Applied Earth Observations and Remote Sensing 5(3):1024-1032.

Yoshikawa, K. and L. D. Hinzman. 2003. Shrinking thermokarst ponds and groundwater dynamics in discontinuous permafrost near Council, Alaska. Permafrost and Periglacial Processes 14(2):151-160.

Yoshikawa, K., C. Leuschen, A. Ikeda, K. Harada, P. Gogineni, P. Hoekstra, L. Hinzman, Y. Sawada, and N. Matsuoka. 2006. Comparison of geophysical investigations for detection of massive ground ice (pingo ice). Journal of Geophysical Research-Planets 111(E6).

Yubao, Q., H. Guo, S. Jiancheng, K. Shichang, J. R. Wang, J. Lemmetyinen, and J. Lingmei. 2010. Analysis between AMSR-E swath brightness temperature and ground snow depth data in winter time over Tibet Plateau, China. Presented at 2010 IEEE International Geoscience and Remote Sensing Symposium (IGARSS).

Zhou, T., P. Shi, J. Luo, and Z. Shao. 2008. Estimation of soil organic carbon based on remote sensing and process model. Frontiers of Forestry in China 3(2):139-141.

Zulueta, R. C., W. C. Oechel, H. W. Loescher, W. T. Lawrence, and K. T. Paw U. 2011. Aircraft-derived regional scale CO_2 fluxes from vegetated drained thaw-lake basins and interstitial tundra on the Arctic Coastal Plain of Alaska. Global Change Biology 17(9):2781-2802.

Appendix A

Abstracts of Workshop Presentations

ASSESSING PERMAFROST EXTENT AND CONDITION FROM REMOTELY SENSED IMAGERY

Larry D. Hinzman, International Arctic Research Center, University of Alaska Fairbanks

Permafrost extent, condition, and processes are specifically identified in the Decadal Survey (NRC, 2007) as an observational need and requirement for understanding climate variability and change. For conditions hidden from direct space view such as permafrost, the NRC has recommended (NRC, 2007, p. 260) that "inferences be drawn from in situ measurements and remotely sensed observations from satellite and suborbital platforms." To date, however, no strategy or NASA missions specifically address the scientific questions surrounding permafrost degradation (NRC, 2007, Table 9.A.1).

We need to develop the remote sensing techniques to produce maps that delineate areas at risk of permafrost degradation and coastal erosion, and produce vulnerability maps for determining safe building locations and provide information where mitigation efforts should be focused to protect Arctic coastal areas. Both permafrost stability and condition can be sensed remotely using surface expression radar, a depth sounding radar, radiometers, airborne CO_2 and CH_4 flux measurements, and ground observations. It has been proven that surface expressions associated with permafrost degradation can be detected and used to infer information about ecological and hydrological systems (e.g., Grosse et al., 2006; Jorgenson et al., 2001; Kääb, 2008; Osterkamp et al., 2009; Stow et al., 2003; Yoshikawa and Hinzman, 2003), yet this has not been done over a large scale and gas fluxes associated with thawing permafrost have not been adequately quantified. A NASA research priority should relate surface expressions of degrading permafrost to the ecological, biological, hydrological, and carbon systems over a large spatial area providing extent and rate quantification of degrading permafrost and gas flux. Attention should focus upon different manifestations of permafrost degradation, including (A) impacts on terrestrial ecosystems and trace gas exchange with the atmosphere; (B) thermokarst topography and lake development/shrinkage; and (C) coastal erosion. It should be possible to sense these manifestations of permafrost degradation respectively using (1) multifrequency coherent radar and radiometers, (2) airborne and ground-based CO_2 and CH_4 flux measurements, (3) LiDAR and hyperspectral imaging, and (4) and year-round ground-based and in situ measurements of permafrost. The size of early thermokarst features is in the range of a few meters (thermokarst pits and sinkholes, thaw slumps, ponds). Growth rates can be several meters (pits and sinkholes) to tens of meters (thaw slumps) per season. Thermokarst lake shore erosion is between 0.25 to more than 7 m/yr depending on lake type, region, and shore configuration.

Surface subsidence is on the order of a few cm to tens of cm per year for very active thermokarst subsidence; tens of centimeters to a few meters for thaw slumps; and tens of centimeters to meters for deep but spatially limited sinkholes (ice wedge degradation). Alterations in soil moisture, soil temperature, and associated vegetation changes resulting from permafrost degradation have profound and interacting effects on fluxes of carbon and energy (Vourlitis et al., 1993). Both vegetation composition and structure change with permafrost degradation due to direct alteration of the soil hydrological and thermal regime in addition to secondary changes in soil nutrients (Christensen et al., 2004; Jorgenson et al., 2001; Stow et al., 2004). Changes in vegetation will affect the rate and amount of above- and below-ground new carbon storage as well as the surface energy balance through changes in albedo, permafrost insulation, and evapotranspiration, which in turn feed back into the soil hydrological and thermal regime. Decomposition of soil organic matter (and its form of carbon release as CO_2 or CH_4) will adjust according to the direct relaxation or enhancement of physiological constraints, the size of the unfrozen organic matter pool, and feedbacks to these factors as well as changes in soil organic matter caused by vegetation.

THE ALASKA SATELLITE FACILITY (ASF): PROVIDING REMOTE SENSING DATA IN SUPPORT OF ARCTIC RESEARCH

Don Atwood, Michigan Tech Research Institute

Rapid ecological change suggests that the Arctic may be a bellwether for the impact of global warming upon more temperate parts of the Earth. This suggests the need for a deeper understanding of baseline processes, environmental drivers, and ecological responses in the high latitudes. Due to the Arctic's vast size, inhospitable conditions, and poor infrastructure, remote sensing will necessarily play an important role in understanding its evolution. The Alaska Satellite Facility (ASF) of the University of Alaska Fairbanks (UAF) is well positioned to support this research. ASF has operated since 1991 as a NASA ground station and archive of satellite data products, with particular focus on synthetic aperture radar (SAR) sensors such as Seasat, JERS-1, ERS-1, ERS-2, RADARSAT-1, and ALOS PALSAR. More recently ASF has become the archive for JPL's Uninhabited Aerial Vehicle SAR (UAVSAR) and Airborne Microwave Observatory of Subcanopy and Subsurface (AirMOSS) data, and will provide data for the upcoming Soil Moisture Active Passive (SMAP) mission. To date, ASF has established an archive of approximately 2 PB of satellite imagery; most of which covers Alaska, western Canada, the Bering Sea, and the Arctic Ocean.

The goal of this talk is to introduce Arctic researchers to the wealth of data and tools that are available through ASF. For example, long-term historical data sets can be useful for visualizing the nature and evolution of North Slope lakes. In a recent project, ERS-1 and -2 SAR data were used to characterize the bathymetry for all North Slope lakes. More recent polarimetric data from ALOS PALSAR and UAVSAR can provide important microwave scattering data, to assist in understanding land cover/land change, as well as the delineation of wetlands. Beginning in 2015, ASF will begin distributing SMAP data which will be extremely useful for understanding permafrost distribution and dynamics. SMAP products include global freeze/thaw and soil moisture maps (updated every 3 to 10 days), as well as imagery from the onboard SAR and radiometer instruments. With an anticipated mission life of 3 years, SMAP will capture the entire annual hydrological cycle as well as chronicle interannual variations attributable to melting permafrost. The presentation will finish with a brief description of how ASF data can be freely acquired for research purposes.

WHAT LIES BENEATH: AIRBORNE ELECTROMAGNETIC METHODS FOR MAPPING SUBSURFACE PERMAFROST AND BUILDING GEOLOGICAL FRAMEWORKS IN COLD REGIONS

Burke J. Minsley, U.S. Geological Survey, Crustal Geophysics and Geochemistry Science Center, Denver, CO

Airborne electromagnetic (AEM) methods play a unique role in the remote sensing of permafrost, and related geological and hydrological environments, because of their ability to map the subsurface from depths of a few meters to several hundred meters below ground. In fact, AEM is not often grouped with more traditional remote sensing technologies, but rather is classified with geophysical methods—a somewhat arbitrary and misinformative distinction. AEM is the only available remote sensing tool that helps to bridge the gap between other airborne and satellite technologies that map surface (or very shallow) features over large areas, and the sparse ground truth information about physical properties at depth from borehole data.

AEM is a decades-old technology borne out of the mineral exploration industry, but has recently seen widespread application in geological and hydrological mapping programs, as well as permafrost and sea-ice thickness studies, that has been facilitated by improvements in instrumentation and processing methods. AEM relies on the physics of electromagnetic induction, as opposed to many other remote sensing modalities that rely on wave propagation, to detect physical properties from the near surface down to several hundred meters below ground. Depth imaging is achieved by acquiring data at different frequencies (several hundred Hz to approximately 100 kHz), where lower frequencies are sensitive to deeper structures. Inductive electromagnetic methods are primarily sensitive to the electrical characteristics of geological materials that, in turn, are a function of properties such as unfrozen water content, lithology, and salinity. A significant challenge in the interpretation of AEM data is to infer the underlying physical properties that result in the mapped distribution of electrical resistivity.

I will introduce the basic instrumentation and methods behind AEM surveying and interpretation, along with examples and ideas of how AEM data can be integrated with other remote sensing products and ground-based measurements for robust, multiscale mapping of permafrost systems. For example, an AEM survey acquired by the USGS in the Yukon Flats of Alaska revealed the subsurface geometry of discontinuous permafrost, and also captured the thermal legacy of the Yukon River lateral migration over the past ~1,000 years, which has been recorded in permafrost. Other AEM surveys acquired by the Alaska Division of Geological and Geophysical Surveys (DGGS) are being interpreted to better constrain permafrost distributions in diverse and complex geological settings. And a recent NSF-supported AEM survey has mapped glacier ice, permafrost, and saline water in portions of Antarctica's dry valleys. These results illustrate the value of AEM data for developing three-dimensional geological and hydrological frameworks of permafrost environments, and the importance of furthering the use of AEM to complement our permafrost remote sensing toolbox.

HOW TO IMPROVE PERMAFROST MODELS USING REMOTE SENSING

Kevin Schaefer (NSIDC), Lin Liu (Stanford), Howard Zebker (Stanford), Andrew Parsekian (Stanford), Elchin Jafarov (NSIDC), Santosh Panda (UAF), Tingjun Zhang (NSIDC)

Improving the representation of permafrost is a key factor in improving the performance of global climate models. We need both in situ and remote sensing data of permafrost characteristics and processes for model validation and parameterizations. Measurements of permafrost temperature, active layer thickness, and other characteristics provide validation or initial values for model prognostic variables. Measurements of landscape changes due to various permafrost processes provide validation data and inputs to model parameterizations. A model parameterization is an observationally based, statistical representation of the large-scale effects of subgrid processes. Models use parameterizations to represent processes that occur on a physical scale that is much smaller than the model resolution, such as wetland dynamics, hydrology, runoff, erosion, fire, insect infestation, snow dynamics, and thermokarst. All of these parameterizations need improvement, particularly parameterizations of thermokarst processes, which are exceedingly rare in current global models.

We recommend a remote sensing strategy designed to improve model parameterizations of permafrost processes. First, you identify surface properties known to reflect large-scale effects of permafrost processes and modify models to simulate these properties. You use remote sensing and in situ data to measure how these properties change over time and develop statistical relationships between the measurements and the effect of a specific process. The model parameterization is the statistical relationships applied to the simulated property to estimate the bulk effects of a permafrost process. To illustrate the strategy, we show how to use interferometric synthetic aperture radar (InSAR) and optical remote sensing data to develop a parameterization of thermokarst lake expansion.

We also recommend expanding the use of geophysical remote sensing techniques to complement current and planned remote sensing capabilities. Geophysical remote sensing includes InSAR, electromagnetic methods, ground-penetrating radar, nuclear magnetic resonance, and related techniques. These and other techniques are widely used on a variety of platforms, including manual deployment, vehicles, aircraft, and satellites. Field measurements taken with such instruments are often lumped with in situ data, but they are, in fact, remote sensing. Geophysical remote sensing often leverages differences in the physical properties of water, ice, and soil, the main ingredients of permafrost. Geophysical techniques have been widely applied to the cryospheric study of glaciers, land ice sheets, and sea ice, but are greatly underutilized in the study of permafrost. We offer a number of potential examples to illustrate the full potential of geophysical remote sensing in understanding key permafrost processes.

SPATIAL AND TEMPORAL RESOLUTION REQUIREMENTS FOR REMOTE OBSERVATION OF PERMAFROST LANDSCAPE DYNAMICS

Guido Grosse,[1] Benjamin Jones,[2] and Vladimir Romanovsky[1]

Permafrost, at regional scales, can be seen as a relatively homogeneous subsurface property—a state of ground temperature defined by areal extent or average vertical thickness. However, at local scales heterogeneity in ground ice content and distribution, soil organic layer thickness, soil stratigraphy, and external factors such as snow and vegetation distribution, micro- and meso-topography, and hydrological framework lead to variability in the response of permafrost landscapes to change. Thus, the scale at which observations are made is important for detecting local disturbances of the ground thermal regime that may lead to thermokarst and other thaw-related landscape features.

Land surface features and processes often allow derivation of information about the local state of permafrost in an area. Identification, mapping, and monitoring of such features/processes with remote sensing can provide access to various subsurface properties and dynamics of near-surface permafrost and the active layer, greatly helping in understanding vulnerabilities and trajectories of change. An important question is what spatial and temporal resolution requirements have to be met by remote sensors to adequately capture local permafrost features and landscape dynamics useful for interpreting the local state and vulnerability of permafrost.

In our presentation we will show examples of permafrost landscape features and processes and discuss what their spatial scales and temporal dynamics are, including thermokarst pond growth, lake and coastal erosion, thaw slump development, peatland collapse, changes in active layer thickness, broad surface subsidence, as well as pingos, ice wedge networks, and small-scale patterned ground. We will differentiate between seasonal versus long-term changes and forward the notion that permafrost change may express both as degradation as well as aggradation.

We (1) will highlight a range of past and current optical remote sensors and their capabilities and limitations in capturing these permafrost landscape dynamics at sufficient spatial and temporal resolution. We will further discuss the application and need for (2) highly accurate and high-resolution digital elevation models, (3) repeat acquisition of elevation and optical remote sensing data sets, and (4) multisensor approaches joining optical and SAR capabilities.

CURRENT STATUS AND FUTURE OF SATELLITE REMOTE SENSING TO BETTER UNDERSTAND ECOLOGICAL IMPACTS TRIGGERED BY CHANGING PERMAFROST

Dara Entekhabi, MIT

The NASA Soil Moisture Active Passive (SMAP) is due to launch in November 2014. The mission provides for global mapping and monitoring of landscape freeze/thaw (FT) status and surface soil moisture conditions. The SMAP Level 2/3 FT product will quantify the predominant frozen or nonfrozen status of the landscape at approximately 3-km resolution and 3-day fidelity. The FT retrievals will be validated to a mean spatial classification accuracy of 80%, sufficient to quantify frozen season constraints to terrestrial water mobility and the potential vegetation growing season

[1]Geophysical Institute, University of Alaska Fairbanks.
[2]Alaska Science Center, U.S. Geological Survey, Anchorage.

over northern (≥45°N) land areas. A SMAP Level 4 carbon (L4_C) product uses the FT retrievals and model value-added surface and root zone soil moisture estimates with other ancillary inputs to quantify net ecosystem CO_2 exchange (NEE), component carbon fluxes, and surface (<10 cm depth) soil organic carbon (SOC) stocks over all global vegetated land areas. The L4_C product also quantifies underlying environmental controls on these processes, including soil moisture and frozen season constraints to productivity and respiration. The L4_C NEE estimates will be validated to an RMSE requirement of 30 g C m^2 yr^{-1} or 1.6 g C m^2 day^{-1}, similar to accuracy levels determined from in situ tower eddy covariance CO_2 flux measurements. The L4_C research outputs include SOC, vegetation productivity, ecosystem respiration, and environmental constraint (EC) metrics clarifying FT and soil moisture–related restrictions to estimated carbon fluxes. These products are designed to clarify how ecosystems, especially in boreal regions, respond to climate anomalies and their capacity to reinforce or mitigate global warming.

The SMAP mission provides for global mapping of soil moisture and landscape FT state dynamics with enhanced L-band (1.26/1.41 GHz) active passive microwave sensitivity to surface soil conditions, and approximate 3-day temporal revisit owing to its 1000- km-wide swath. The radar resolution is better than 3 km in the outer 70% of the swath and away from the satellite track at nadir. The SMAP satellite is in a polar orbit which results in considerable overlap of swaths in northern latitudes. By combining the far outer edge of overlapping swaths each day, it is possible to construct a daily L-band radar mapping of northern latitudes at 1 km resolution. The measurements are valuable to monitoring the land and sea cryosphere regardless of clouds, weather, and solar illumination. We propose a community effort to produce an all-Alaska daily 1-km-resolution L-band radar backscatter cross-section product based on Level 1 SMAP files at the Alaska Satellite Facility (NASA-designated Distributed Active Archive Center for SMAP radar data).

AIRBORNE REMOTE SENSING CAPABILITIES TO UNDERSTAND ECOLOGICAL IMPACTS TRIGGERED BY CHANGING PERMAFROST

Charles Miller, Jet Propulsion Laboratory, California Institute of Technology

The presentation includes a survey of airborne instruments and techniques for tracking high-latitude ecology and atmospheric carbon dioxide and methane. There is also a brief survey of existing airborne assets for permafrost characterization, including both radar and electromagnetic (EM) methods.

There are two significant ecological impacts of permafrost change. One is a "greening Arctic," which results in a change in species and range of vegetation cover, and an increase in carbon uptake. Second is an increase in carbon dioxide and methane emissions from mobilized ancient carbon. Carbon dioxide/methane fractioning depends critically on changes in hydrology. The future carbon budget of the northern high latitudes depends on the (im)balance that a changing permafrost imposes.

Airborne remote sensing also offers the potential for multisensor observations that can bring unique insights into the ecosystem processes and properties. Asner et al. (2007) pioneered the fusion of high-fidelity visible/VIS-SWIR hyperspectral imaging spectrometer data with scanning, waveform light detection and ranging (wLiDAR) data, along with an integrated navigation and data processing approach. This is a quantum leap beyond BOREAS ecosystem remote sensing. It retrieves information on vegetation canopy structure, vegetation biochemistry, vegetation biophysical properties, and the ecosystem response. It will be available on a regular basis in Alaska via NEON (National Ecological Observatory Network) beginning around 2014.

Solar induced chlorophyll fluorescence (SIF) can be measured through high-spectral-resolution remote sensing in the 690-770 nm region. SIF is directly related to photosynthetic activity, although the exact functional relationship of SIF to GPP (Gross Primary Production) is currently debated. SIF is measured by airborne sensors (FLEX Simulator, CARVE FTS, MAMAP) and satellite instruments (GOSAT, OCO-2, OCO-3, FLEX [proposed], CarbonSat [proposed]).

It is also possible to measure the total column carbon dioxide and methane using high-spectral-resolution remote sensing in the 1650-2400 nm region. Airborne sensors with this capability include CARVE FTS and MAMAP. Satellites that can measure carbon dioxide include GOSAT, OCO-2, OCO-3, and CarbonSat (proposed). Satellites that can measure methane include SCIAMACHY (no longer available), GOSAT, CarbonSat (proposed), and the Sentinel 5 precursor. It is important to note that **NASA currently has no plans for a space-based mission to measure methane over the high latitudes.**

The presentation also includes several examples of airborne methods that can be used for permafrost characterization: (1) the AirMOSS Flight System; (2) the Boreal Ecosystem Research Monitoring Sites (BERMS); (3) mapping the average seasonal subsidence between 1992 and 2000 near Prudhoe Bay, Alaska, by utilizing a time series of 14 interferograms from the ERS satellite; (4) mapping the average active layer thickness (ALT) between 1992 and 2000 near Prudhoe Bay by converting the seasonal subsidence to melted water and assuming a vertical water distribution; (5) resistivity cross sections in the Yukon-Flats Region; and (6) three-dimensional mapping with HEM (helicopter electromagnetic).

A GEOBOTANICAL PERSPECTIVE: MONITORING ARCTIC PERMAFROST AND ECOSYSTEM CHANGE USING REMOTE SENSING, GIS, AND PLOT-BASED STUDIES

D.A. Walker Institute of Arctic Biology, University of Alaska Fairbanks, USA

Integrated mapping approaches for the Arctic have been evolving. Since beginning in 1969, the Alaska Geobotany Center (AGC) has made vegetation maps using traditional photo-interpretive methods, satellite sensors, and plot-based interdisciplinary research along environmental gradients. These maps have a myriad of applications to permafrost and global-change research. In this talk I discuss four points that I see as essential for developing a comprehensive interdisciplinary approach to use remote sensing to examine Arctic change:

Spatial hierarchy of databases: Databases are needed for answering questions at plot to planet scales. This necessarily requires consistent approaches for visualizing and coloring the maps so that they make intuitive sense across scales. Recent availability of very-high-resolution satellite imagery promises to revolutionize interpretation of changes to permafrost patterned-ground landscapes.

Circumpolar databases: A circumpolar examination of permafrost and environmental change requires pan-Arctic spatial databases. Such databases require a high level of synthesis and international coordination.

Long-term data sets: Time series of ground observations need to complement time series of remote sensing images and detailed mapping.

Integrated mapping studies: Mapping should integrate as much geoecological information from different disciplines as possible into single databases along with historical natural geoecological changes and anthropogenic changes. Examples include the Integrated Terrain Unit Mapping approach developed by ESRI Inc. and the "integrated geoecological and historical change maps" (IGHCM) that were developed to simultaneously examining historical changes caused by dynamic permafrost landscapes and those caused by expanding networks of oil field infrastructure.

The AGC and the Geographic Information Network of Alaska (GINA) are in the process of developing an Arctic Alaska Geoecological Atlas for NASA's planned Arctic Boreal Vulnerability Experiment (ABoVE) that will draw on the principles discussed above (http://above.nasa.gov).

THE CONTRIBUTION OF SPACEBORNE SYNTHETIC APERTURE RADAR SENSORS TO PERMAFROST RESEARCH

Franz J. Meyer, Associate Professor, University of Alaska Fairbanks, AK
Paul A. Rosen, Jet Propulsion Laboratory, Pasadena, CA

In the recent decade, data synthetic aperture radar (SAR) sensors have been shown to have great potential for observing the Arctic. This is in large part due to two advantageous characteristics of SAR data: (1) As an active sensor, SAR systems can observe the ground independent of weather and illumination conditions and are as such the only systems that can reliably provide 24/7 observations and (2) in addition to image information, SAR phase observations can provide measurements of surface dynamics, which can be used to indicate surface change and measure centimeter-scale surface deformation. Owing to these benefits, SAR data have the potential to provide information on two major processes in permafrost regions that are relevant for understanding climate change impacts in northern high-latitude environments: short-term, seasonal dynamics of the active layer located above permafrost, and long-term multiannual changes in permafrost extent.

We will show in theory and examples that current and future SAR missions can provide information on several key parameters for seasonal active layer (freeze/thaw, subsidence and heave, deep soil moisture) and long-term permafrost dynamics (subsidence, lateral movements). We will furthermore particularly highlight the planned capabilities of an upcoming proposed NASA L-band SAR mission that can provide permafrost-related information at high spatial and temporal resolution and at accuracy levels that may lead to a substantial improvement of our understanding of panarctic active layer dynamics and permafrost thaw. We will summarize the mission's predicted measurement characteristics by highlighting its high temporal and spatial sampling, its global observation strategy, and its predicted performance in measuring surface dynamics. Based on the system's proposed measurement capabilities, we claim that this future L-band mission will allow for a spatially explicit assessment of regional to global impacts of permafrost dynamics on hydrology, carbon cycling, and northern ecosystem character and functioning.

MICROWAVE REMOTE SENSING UTILITY FOR DOCUMENTING ENVIRONMENTAL CHANGE IN PERMAFROST LANDSCAPES

John S. Kimball, Flathead Lake Biological Station, Division of Biological Sciences, The University of Montana

This presentation highlights the potential utility of satellite active and passive microwave remote sensing for regional monitoring of physical attributes related to soil active layer dynamics and permafrost in the boreal/arctic. Relative strengths and limitations of current global satellite records and the potential utility of upcoming NASA Earth missions are discussed. Potential research gaps are identified and recommendations for improving the relevance of these observations for permafrost landscapes are presented.

Satellite active and passive microwave remote sensing at lower frequencies (~≤37 GHz) has strong utility for mapping and monitoring of physical land parameters relevant to soil active layer dynamics in permafrost landscapes. Satellite passive microwave sensors detect natural microwave emissions of the land surface, while the associated brightness temperature (T_b) retrievals are strongly sensitive to surface moisture and temperature through their effect on surface dielectric properties. Because only a small portion of Earth's energy emissions are at lower microwave frequencies, the T_b retrievals have generally coarse (~12-60 km) spatial resolution to enhance the sensor signal-to-noise ratio. In contrast, active microwave sensors (radars) provide their own land surface illumination source, enabling finer-spatial-scale retrievals with a larger signal-to-noise ratio. The sensitivity to soil attributes is strongly dependent on microwave frequency and land surface conditions. Lower frequencies (e.g., L/P band) have generally greater potential soil active layer sensitivity, while the relative depth of direct soil sensitivity is inversely proportional to the moisture content in soil and overlying snow and vegetation layers. The relative insensitivity of microwaves to solar illumination and atmosphere aerosols, including clouds and smoke, and the converging orbital swaths of polar orbiting sensors enable daily or better temporal fidelity over northern (≥50°N) land areas from operational global satellites, while finer-scale synthetic aperture radar (SAR) sensors have more limited spatial and temporal coverage. These attributes have been exploited for monitoring a range of physical parameters, including surface soil moisture and temperature, landscape freeze/thaw dynamics, open water inundation, snow cover, vegetation biomass, and terrain structure. Similar satellite microwave retrievals from overlapping operational sensor records have also enabled the development of relatively long-term records documenting recent (up to 30+ year) environmental changes with relatively high precision.

New global satellite missions are coming online

that enable new observations of biophysical attributes in permafrost landscapes. The NASA SMAP (Soil Moisture Active Passive) mission has a projected launch in mid-2014 and will provide enhanced L-band (1.26/1.4 GHz) sensitivity to surface soil moisture and landscape freeze/thaw dynamics with regular global monitoring at moderate (1-9 km) spatial resolution and 1-3 day temporal repeat. Model-enhanced (Level 4) products are also planned that will provide daily estimates of soil profile ($\leq$1 m depth) moisture and thermal conditions, surface soil organic carbon (SOC) stocks, and terrestrial carbon (CO_2) fluxes. However, a number of limitations remain for applying existing and planned satellite sensor records for monitoring environmental change in permafrost landscapes. These limitations include a lack of finer resolution (e.g., 10-100 m) monitoring approaching the scale of landscape variability in permafrost attributes. Other limitations include loss of direct sensitivity to soil attributes under higher vegetation biomass and potentially complex processing required for extracting meaningful land parameter information from lower order microwave retrievals. Various methods have been developed for regional downscaling of satellite observations that may enhance utility of these data for permafrost. These techniques include spatial resolution enhancement techniques applied to overlapping T_b and radar backscatter orbital swath data, and empirical modeling and data fusion techniques using synergistic multiscale and multisensor remote sensing, and other ancillary data for estimating finer-spatial-scale attributes. While these techniques have been widely used for other areas, there is still a general paucity of regional applications of these techniques in permafrost landscapes.

New airborne assets have become available that can be used to inform algorithm development and regional downscaling efforts. These assets include the NASA AirMOSS and UAVSAR sensors, which provide for finer-scale L/P-band SAR retrievals. The JAXA JERS-1 SAR and ALOS PALSAR sensor records provide similar fine-scale (~10 to 100 m resolution) L-band satellite SAR data, which have been used to investigate sub-grid-scale landscape freeze/thaw heterogeneity. These initial studies indicate potential utility for quantifying scaling behavior in similar, coarser-scale satellite retrievals that can be used to inform regional downscaling efforts. Similar L-band SAR data will be provided by ALOS-2, which has a projected launch in late 2013. However, user access to these data may be severely constrained by limited access and cost-per-scene data use restrictions, and may require significant NASA investment in potential data buys or data use agreements to secure unrestricted access to these data.

Recent studies using AirSAR and theoretical microwave radiative transfer modeling indicate potential utility for combined L/P-band SAR remote sensing for direct retrievals of soil active layer development in lower biomass (e.g., tundra) areas, while reduced soil sensitivity under higher biomass (e.g., boreal forest) conditions may constrain direct soil retrievals in the boreal zone. However, these constraints may be mitigated by using additional vegetation biomass structure information provided by LiDAR and optical sensors. New investments in coordinated satellite and airborne remote sensing and detailed ground network measurements are needed for further algorithm development and validation (Cal/Val) efforts to fully develop and demonstrate these capabilities.

Several near-term (next 3-5 years) boreal-Arctic field campaigns are under development, including limited campaigns supporting post-launch sensor and product Cal/Val activities for SMAP, OCO-2 (Orbiting Carbon Observatory), and a more extensive NASA-led Arctic Boreal Vulnerability Experiment (ABoVE). These campaigns will involve coordinated satellite, airborne, and field-based measurements, with potential focus on physical parameters directly relevant to permafrost attributes and processes. These activities provide opportunities for testing sensors and developing retrieval algorithms specific to permafrost landscapes. Finally, a number of agency and international efforts are under way or in the planning stages that focus on improving monitoring capabilities and understanding of environmental change in permafrost landscapes. Collaborative partnerships among these agencies and efforts should be encouraged for scoping and communicating joint needs, developing new satellite missions and field campaigns, and the free exchange of data.

POTENTIAL OF HYPERSPECTRAL REMOTE SENSING FOR MAPPING PERMAFROST FEATURES AND ASSOCIATED BIOPHYSICAL VARIABLES

Anupma Prakash, Jordi Cristóbal, Christian Haselwimmer, and Don Hampton, Geophysical Institute, University of Alaska, Fairbanks

Hyperspectral remote sensing, also known as imaging spectroscopy, has potential to make a significant contribution in permafrost research, by providing a tool to map associated biophysical variables with unprecedented detail. A large body of literature exists that documents the success of imaging spectroscopy in identifying and mapping plant species and plant functional types. However, literature on the direct use of imaging spectroscopy for mapping permafrost features or linking permafrost distribution with biophysical variables identified and mapped by imaging spectroscopy is at best limited.

NASA's planned Hyperspectral Infrared Imager (HyspIRI) mission is designed to quantitatively study the Earth's terrestrial biosphere, identify vegetation species and functional types, and provide benchmark mapping against which future changes can be assessed. With two imaging spectrometers, one in the 380 to 2500 nm visible shortwave infrared (VSWIR) region, and the other in the 3 to 12 µm thermal infrared (TIR) region, providing 60 m spatial resolution data and a near global coverage every 19 days (for VSWIR)/5 days (for TIR), the instrument will support a broad spectrum of carbon and water cycle and ecosystem studies. Permafrost research will particularly benefit from HyspIRI's VSWIR instrument that will provide a means for superior identification and classification of Arctic and sub-Arctic vegetation. The potential for better characterizing the mosses and shrubs associated with permafrost landscape will be a distinct advantage. The TIR instrument will allow us to better map energy fluxes and other related biophysical variables (such as soil moisture or air temperature) at medium spatial resolution.

At the University of Alaska Fairbanks (UAF) we have invested in two areas that will support permafrost research: (i) Through a recent grant from the National Science Foundation Major Research Instrumentation (MRI) program, UAF is in the initial stages of developing an in-state capability (The University of Alaska Fairbanks Hyperspectral Imaging Laboratory—UAF HyLab) for airborne and ground-based imaging spectroscopy based around the acquisition of commercial HySpex visible and shortwave infrared (0.4-2.5 µm) hyperspectral systems. The capability for routine acquisition of new imaging spectroscopy data sets over Alaskan study sites will provide a tremendous boost to the permafrost remote sensing, and for studying ecosystem composition and change. (ii) Through a NASA EPSCoR grant, UAF has established two field sites, one in the interior Alaska boreal forest setting and another in the deciduous forest setting, which include a suite of ground-based instrumentations collecting data on surface energy flux, ground heat flux, and other essential climate variables. These field sites, that can serve as calibration and validation (CalVal) sites for satellite missions, are being used to scale observations from field scale to satellite scale.

Appendix B

Workshop Agenda and Participant List

Opportunities to Use Remote Sensing in Understanding Permafrost and Related Ecological Characteristics

Workshop Agenda
October 8-9, 2013
University of Alaska, Fairbanks
International Arctic Research Center
Room 501
Fairbanks, AK

WORKSHOP GOALS:

Permafrost thaw stands to have wide-ranging impacts, such as the draining and drying of the tundra, erosion of riverbanks and coastline, and destabilization of infrastructure (roads, airports, buildings, etc.), and including potential implications for ecosystems and the carbon cycle in the high latitudes. The goal of this workshop is to explore opportunities for using remote sensing to advance our understanding of permafrost status and trends and the impacts of permafrost change, especially on ecosystems and the carbon cycle in the high latitudes.

Attendees at the workshop will address questions such as how remote sensing might be used in innovative ways, how it might enhance our ability to document long-term trends, whether it is possible to integrate remote sensing products with the ground-based observations and assimilate them into advanced Arctic system models, what are the expectations of the quality and spatial and temporal resolution possible through such approaches, and what prototype sensors (e.g., the airborne UAVSAR, AIRSWOT (InSAR) and MABEL (LiDAR), IceBridge) are available and might be used for detailed ground calibration of permafrost/high-latitude carbon cycle studies?

The workshop discussions are designed to encourage attendees to articulate gaps in current understanding and potential opportunities to harness remote sensing techniques to better understand permafrost, permafrost change, and implications for ecosystems in permafrost areas.

Shuttle service will be provided for both days to and from IARC

Tuesday, October 8

7:30 a.m. *Shuttle departs from the Westmark Hotel*

8:00 a.m. *Breakfast*

8:30 a.m. Goals and objectives of the workshop
Prasad Gogineni, University of Kansas
Vladimir Romanovsky, University of Alaska Fairbanks, *Co-Chairs*

8:50 a.m. Setting the stage: Lessons from ESA DUE
Claude Duguay, University of Waterloo

SESSION 1

Moderator: Vladimir Romanov

Usign remote sensing to better understand permafrost properties (distribution, ice content, thermal state, active layer thickness, etc.) and recent changes in permafrost.

9:20 a.m. Permafrost—what is needed?
Larry Hinzman, UAF

9:50 a.m. Satellite—current status and future
Don Atwood, Alaska Satellite Facility

10:20 a.m. *Break*

10:45 a.m. Airborne—current status and future
Burke Minsley, USGS

11:15 a.m. In situ
How to integrate remote sensing with in situ measurements and modeling/reanalKevin Schaefer, NSIDC

11:45 a.m. Discussion

12:30 p.m. *Lunch*

SESSION 2

Moderator: Torre Jorgenson

Using remote sensing to measure the biophysical and/or ecological characteristics to quantify permafrost properties (hydrological changes including lake dynamics, surface heave/subsidence, thermokarst development, thermal erosion, slope instability, changes in micro-topography, changes in vegetation, etc).

1:30 p.m Permafrost—what is needed?
Guido Grosse, UAF

2:00 p.m Satellite—current status and future
Dara Entekhabi, MIT

2:30 p.m. Airborne—current status and future
Chip Miller, JPL

3:00 p.m. *Break*

3:30 p.m. In situ
How to integrate remote sensing with in situ measurements and modeling/reanalysis?
Skip Walker, UAF

4:00 p.m. Discussion

4:45 p.m. Plan for tomorrow
Prasad Gogineni
Vladimir Romanovsky

5:00 p.m. *Adjourn*

5:15 p.m. *Shuttle departs for Westmark Hotel*

Wednesday, October 9

7:30 a.m. *Shuttle departs from the Westmark Hotel*

8:00 a.m. *Breakfast*

SESSION 3

Moderator: Jessie Cherry

What are the major gaps and what is needed to enable remote sensing to make further progress in the above areas? What are the new possibilities of new sensors and planned missions and what changes can be made in the future NASA mission to address these questions?

8:30 a.m. Presentations from remote sensing researchers
Franz Meyer, UAF
John Kimball, University of Montana
Anupma Prakash, UAF

10:00 a.m. **First Breakout Group Session:** Breakouts focusing on compiling a table that captures what is currently possible to measure using remote sensing to study permafrost and what is needed and/or possible for the future. *See Tab D in the briefing book for more details.*

12:15 p.m. *Lunch*

1:15 p.m. Report Back

2:30 p.m. **Second Breakout Group Session**: Breakouts focusing on the following questions. *See Tab D in the briefing book for more details.*

1. How can we establish a baseline that would be most valuable to documenting ongoing change? What sensors are best suited to this need? How might they be used to provide complementary information?
2. How can remote sensing be used in innovative ways and how can it enhance our ability to document long-term trends?
3. What prototype sensors, such as the airborne UAVSAR, AIRSWOT (InSAR) and MABEL (LiDAR), AirMOSS (P-band radar), and IceBridge, are available and could be used with detailed ground calibration and validation for permafrost studies?
4. How can remote sensing products be best integrated with ground-based observations and assimilated into advanced Arctic system models and permafrost models?

4:00 p.m. Report Back

5:15 p.m. Wrap-up and Final Remarks
Prasad Gogineni
Vladimir Romanovsky

5:30 p.m. *Workshop adjourns*

5:45 p.m. *Shuttle departs for Westmark Hotel*

PARTICIPANT LIST

Name	Affiliation
Michel Allard	Université Laval
Don Atwood	Alaska Satellite Facility
Andrew Balser	University of Alaska Fairbanks
Annett Bartsch	Vienna University of Technology
Breck Bowden	University of Vermont
Jessica Cherry*	University of Alaska Fairbanks
Jordi Cristobal	University of Alaska Fairbanks
Claude Duguay*	University of Waterloo
Dara Entekhabi	Massachusetts Institute of Technology
Gerald Frost	ABR, Inc. Environmental Research & Services
Scott Goetz*	Woods Hole Research Center
Prasad Gogineni*	University of Kansas
Santonu Goswami	Oak Ridge National Laboratory
Guido Grosse	University of Alaska Fairbanks
Christian Haselwimmer	University of Alaska Fairbanks
Tom Heinrichs	University of Alaska Fairbanks
Larry Hinzman	University of Alaska Fairbanks
Forrest Hoffmann	Oak Ridge National Lab
Hiroki Ikawa	University of Alaska Fairbanks
Torre Jorgenson*	Alaska Ecoscience
John Kimball	University of Montana
Ray Kreig	RA Kreig & Associates
Mark Lara	University of Alaska, Fairbanks
Tom Rune Lauknes	Northern Research Institute
Mark Lara	University of Alaska Fairbanks
Philip Martin	U.S. Fish and Wildlife Service
Kyle McDonald	City University of New York
Franz Meyer	University of Alaska Fairbanks
Chip Miller	Jet Propulsion Laboratory
Burke Minsley	U.S. Geological Survey
Mahta Moghaddam*	University of Southern California
Bruce Molnia	U.S. Geological Survey
Reginald Muskett	University of Alaska Fairbanks
Dan Nossar	University of Alaska Fairbanks
Santosh Panda	University of Alaska Fairbanks
Tamlin Pavelsky	University of North Carolina
Anupma Prakash	University of Alaska Fairbanks
Bernhard Rabus	MDA Corporation
Vladimir Romanovsky*	University of Alaska Fairbanks
Kevin Schaefer	National Snow and Ice Data Center
Ted Schuur	University of Florida
Alexei Skurikhin	Los Alamos National Laboratory

David Swanson	National Park Service
Isabella Velicogna	University of California
Skip Walker	University of Alaska Fairbanks
Simon Yueh	Jet Propulsion Laboratory

*committee member

Appendix C

Statement of Task

Permafrost thaw stands to have wide-ranging impacts, such as the draining and drying of the tundra, erosion of riverbanks and coastline, and destabilization of infrastructure (roads, airports, buildings, etc.), including potential implications for ecosystems and the carbon cycle in the high latitudes. Under the auspices of the National Research Council, a committee of experts will plan a workshop to explore opportunities for using remote sensing to advance our understanding of permafrost status and trends and the impacts of permafrost change, especially on ecosystems and the carbon cycle in the high latitudes.

Attendees at the workshop would address questions such as how remote sensing might be used in innovative ways, how it might enhance our ability to document long-term trends, whether it is possible to integrate remote sensing products with the ground-based observations and assimilate them into advanced Arctic system models, what are the expectations of the quality and spatial and temporal resolution possible through such approaches, and what prototype sensors (e.g., the airborne UAVSAR, AIRSWOT (InSAR) and MABEL (LiDAR), IceBridge) are available and might be used for detailed ground calibration of permafrost/high-latitude carbon cycle studies?

The workshop will bring together experts from the remote sensing community with permafrost and ecosystem scientists. The workshop discussions will be designed to encourage attendees to articulate gaps in current understanding and potential opportunities to harness remote sensing techniques to better understand permafrost, permafrost change, and implications for ecosystems in permafrost areas.

Appendix D

Committee Biosketches

Prasad Gogineni is a Deane E. Ackers Distinguished Professor in the Electrical Engineering and Computer Science Department at the University of Kansas and Director of the NSF Science and Technology Center for Remote Sensing of Ice Sheets (CReSIS). He is an IEEE Fellow and served as Manager of NASA's Polar Program Office from 1997 to 1999. Dr. Gogineni received the Louise Byrd Graduate Educator Award at the University of Kansas and was a Fulbright Senior Scholar at the University of Tasmania in 2002. He has been involved with radar sounding and imaging of ice sheets for more than 15 years and contributed to the first successful demonstration of SAR imaging of the ice bed through more than 3-km-thick ice. Dr. Gogineni has authored or co-authored more than 100 archival journal publications and more than 200 technical reports and conference presentations.

Vladimir E. Romanovsky is a Professor of Geophysics in the Geophysical Institute and Geology and Geophysics Department with the University of Alaska, Fairbanks. He is involved in research in the field of permafrost geophysics, with particular emphasis on the ground thermal regime, active layer and permafrost processes, and the relationships between permafrost, hydrology, biota, and climate. He is also dealing with the scientific and practical aspects of environmental and engineering problems involving ice and permafrost, subsea permafrost, seasonally frozen ground, and seasonal snow cover. Dr. Romanovsky is also interested in the improvement of mathematical methods (analytical and numerical modeling) in geology and geophysics. He received his Ph.D. in geology from Moscow State University in 1982 and his Ph.D. in geophysics from University of Alaska Fairbanks in 1996.

Jessica Cherry is a Research Assistant Professor at the University of Alaska, Fairbanks. Her research interests include arctic hydrology and climate, large-scale snow physics, land-atmosphere interaction on synoptic and longer time scales, frozen ground, and water resources and economics. She received her Ph.D. in climate science and hydrology from Columbia University in 2006.

Claude Duguay is a Professor at the University of Waterloo. His main research interests are in remote sensing and modeling of cold regions with the intent of deepening our knowledge, understanding, and predictive capabilities of lake/land-atmosphere/climate interactions. Some of his current areas of interest include the development of satellite-based lake and permafrost-related products, the response of lakes to contemporary and future (projected) climate conditions, the role of lakes in weather and climate, and improvement of the representation of cryospheric processes in lake model schemes as implemented in numerical weather prediction and climate models. Dr. Duguay received his Ph.D. at the University of Waterloo in 1989.

Scott Goetz is the Deputy Director and a Senior Scientist at the Woods Hole Research Center. His research focuses on analysis of environmental change, including monitoring and modeling links between climate and land use change of various types (e.g., urbanization, fire disturbance, deforestation) and their combined influence on biological diversity, water quality, and ecosystem carbon cycling. Dr. Goetz received his Ph.D. from the University of Maryland in 1996.

M. Torre Jorgenson is the of owner of Alaska Ecoscience, a small business in Fairbanks, Alaska, dedicated to research on Alaska's changing landscapes. He also is affiliate faculty with the Departments of Biology and Wildlife, Geology and Geophysics, and Civil Engineering at the University of Alaska, Fairbanks, and is a past president of the U.S. Permafrost Association. Previously, he was a Senior Scientist with ABR, Inc., for 24 years. He has worked on ecology and geomorphology studies throughout Alaska for more than 30 years, focusing on vegetation-soil-permafrost interactions and ecological impacts of human activities. A primary focus has been ecological land classification and terrain mapping, coastal studies, and soil carbon/permafrost dynamics throughout Alaska. He was a steering committee member for the international Arctic Coastal Dynamics project and is a U.S. mapping team member for the Circumboreal Vegetation Mapping project. He has conducted numerous studies on oilfield impacts and land restoration in northern Alaska over several decades. Current projects include characterizing and mapping permafrost in northern Alaska, mapping permafrost and soil carbon in the Yukon River Basin, quantifying changes in hydrology and soil carbon after permafrost thaw in central Alaska, assessing effects of climate change on permafrost and landscapes on military lands, assessing effects of glacial thermokarst in northern Alaska, modeling changes in habitats from climate warming in northwest Alaska, and monitoring long-term ecological changes on the Yukon Kuskokwim Delta since 1994.

Mahta Moghaddam is a Professor at the University of Southern California. Her research interests include radar systems, remote sensing, environmental sensing, medical imaging, focused microwave therapy systems, inverse scattering, and subsurface sensing. Dr. Moghaddam has introduced innovative approaches and algorithms for quantitative interpretation of multichannel radar imagery based on analytical inverse scattering techniques applied to complex and random media. She has also developed quantitative approaches for multisensor data fusion by combining radar and optical remote sensing data for nonlinear estimation of vegetation and surface parameters. She has led the development of new radar instrument and measurement technologies for subsurface and subcanopy characterization. Dr. Moghaddam received her Ph.D. in electrical and computer engineering from the University of Illinois, Urbana-Champaign, in 1991.

Appendix E

Acronyms and Initialisms

ABoVE Arctic-Boreal Vulnerability Experiment
AEM Airborne Electromagnetic
AirMOSS Airborne Microwave Observatory of Subcanopy and Subsurface
AIRS Atmospheric Infrared Sounder
AIRSAR Airborne Synthetic Aperture Radar
AirSWOT Air Surface Water and Ocean Topography
ALOS-2 Advanced Land Observing Satellite-2
ALT Active Layer Thickness
AMSR Advanced Microwave Scanning Radiometer
ASCAT Advanced Scatterometer
ASTER Advanced Spaceborne Thermal Emission and Reflection Radiometer
ATLAS Advanced Topographic Laser Altimeter System
AVHRR Advanced Very High Resolution Radiometer
AVIRIS Airborne Visible/Infrared Imaging Spectrometer

CALM Circumpolar Active Layer Monitoring
CASI Compact Airborne Spectrographic Imager
CIR Color Infrared
$CoReH_2O$ Cold Regions Hydrology
CrIS Cross-track Infrared Sounder

DEM Digital Elevation Model
DOE U.S. Department of Energy
DOI U.S. Department of the Interior
DUE Data User Element

ECV Essential Climate Variable
EM Electromagnetic Spectrum
ENVISAT Environmental Satellite
EO-1 Earth Observing 1
ESA European Space Agency
EU European Union

FLEX Fluorescence Explorer

GCOM-C Global Change Observation Mission (Climate)
GCOM-W Global Change Observation Mission (Water)
GCOS Global Climate Observing System
GIS Geographic Information System
GOES Geostationary Operational Environmental Satellite
GOSAT Greenhouse Gases Observing Satellite
GPM Global Precipitation Measurement
GPR Ground-Penetrating Radar
GRACE Gravity Recovery and Climate Experiment
GRACE-FO Gravity Recovery and Climate Experiment-Follow-On

HRV High-Resolution Visible

HyspIRI	Hyperspectral Infrared Imager
IEM	Integrated Ecosystem Model
IGOS	Integrated Global Observing Strategy
InSAR	Interferometric Synthetic Aperture Radar
IR	Infrared
JAXA	Japan Aerospace Exploration Agency
LAI	Leaf Area Index
LiDAR	Light Detection and Ranging
LST	Land Surface Temperature
MABEL	Multiple Altimeter Beam Experimental LiDAR
MERRA	Modern Era-Retrospective Analysis for Research and Applications
MLA	Mercury Laser Altimeter
MODIS	Moderate Resolution Imaging Spectroradiometer
NARR	North American Regional Reanalysis
NASA	National Aeronautics and Space Administration
NDVI	Normalized Difference Vegetation Index
NGEE	Next Generation Ecosystem Experiments
NIR	Near Infrared
OCO	Orbiting Carbon Observatory
OIB	Operation IceBridge
PALSAR	Phased Array L-band Synthetic Aperture Radar
POES	Polar Operational Environmental Satellite
QPE	Quantitative Precipitation Estimate
RF	Radio Frequency
RGB	Red/Green/Blue
SAR	Synthetic Aperture Radar
SCA	Snow-Covered Area
SIR-C/X-SAR	Spaceborne Imaging Radar-C/X-Band Synthetic Aperture Radar
SMAP	Soil Moisture Active/Passive
Snotel	Snowpack Telemetry
SNPP	Suomi National Polar-orbiting Partnership
SPOT	Satellite Pour l'Observation de la Terre
SRTM	Shuttle Radar Topography Mission
SSM	Surface Soil Moisture
SSM/I	Special Sensor Microwave Imager
SWE	Snow Water Equivalent
SWOT	Surface Water Ocean Topography
TIR	Thermal Infrared
TWS	Terrestrial Water Storage
UAV	Unmanned Aerial Vehicle
UAVSAR	Unmanned Aerial Vehicle Synthetic Aperture Radar
UHF	Ultra High Frequency
USGS	U.S. Geological Survey
VHF	Very High Frequency
VIS-IR	Visible Infrared
VLF	Very Low Frequency
VSWIR	Visible Shortwave Infrared
WMO	World Meteorological Organization